AF595036

Lab-on-PCB Devices

Lab-on-PCB Devices

Editor

Francisco Perdigones

MDPI • Basel • Beijing • Wuhan • Barcelona • Belgrade • Manchester • Tokyo • Cluj • Tianjin

Editor
Francisco Perdigones
Deparment Electronic
Engineering
University of Seville
Seville
Spain

Editorial Office
MDPI
St. Alban-Anlage 66
4052 Basel, Switzerland

This is a reprint of articles from the Special Issue published online in the open access journal *Micromachines* (ISSN 2072-666X) (available at: www.mdpi.com/journal/micromachines/special_issues/Lab_PCB).

For citation purposes, cite each article independently as indicated on the article page online and as indicated below:

LastName, A.A.; LastName, B.B.; LastName, C.C. Article Title. *Journal Name* **Year**, *Volume Number*, Page Range.

ISBN 978-3-0365-4702-2 (Hbk)
ISBN 978-3-0365-4701-5 (PDF)

Contents

About the Editor

Francisco Perdigones

Francisco Perdigones (associate professor), Electronic Devices and Systems Group (DISEL), Electronic Engineering Department, Higher Technical School of Engineering, University of Seville, 41092 Seville, Spain. His research interests include integration of sensors and/or actuators on lab-on-PCB platforms; marketable lab-on-PCB devices; biomedical and chemical applications and PCB-MEMS devices. He is working on several research projects, for example the development of biomedical marketable microdevices for DNA amplification (PCR and LAMP), the design on cell-electronic interfaces, and the fabrication of devices to understand and improve the retinitis pigmetosa disease using low cost microelectrode arrays.

Preface to "Lab-on-PCB Devices"

Lab-on-PCB (LoP) has been the subject of increasing research in recent years. These kinds of devices emerge as an interesting evolution of "lab on a chip" and "PCB-MEMS". They share many properties with lab-on-chip devices and microfluidics, namely, rapid response time and small fluid volume for samples and reagents. In addition, lab-on-PCB is especially interesting due to the integration of electronics, microfluidics, sensors and actuators in a single platform. Beyond this integration, the interest lies in the commercial availability of the Printed Circuit Boards with reasonable dimensions and tolerances at low cost. This fact makes LoP devices an attractive option from the market point of view. However, lab-on-PCB is far from being a robust technology. Unlike electronic microchips, the development of LoP devices covers many fields (electronics, materials, biology, medicine, fluid mechanics, etc.) and therefore they require a highly multidisciplinary R&D group. In addition, they are lacking in standardization for both design and end-user interfaces. Futhermore, the core of LoP devices is based on the integration of sensors, biosensors and actuators. The actuators are intended to move the samples through the LoP device, and for conditioning the samples, for example using integrated microheaters.

This work is addressed to researchers interested in developing mass-produced biomedical devices and microdevices. In addition, this is a good guide for beginners to learn about Lab-on-PCB devices and PCB-MEMS thanks to the review papers and the rest of the contributions as current interesting examples.

I would like to thank all the authors for submitting their papers to the Special Issue "Lab on PCB Devices". I also thank all the reviewers, editors and the MDPI staff for taking their time to improve the quality of the papers.

Francisco Perdigones
Editor

MDPI

Editorial

Editorial for the Special Issue on Lab-on-PCB Devices

Francisco Perdigones 

Electronic Devices and Systems Group (DISEL), Electronic Engineering Department, Higher Technical School of Engineering, University of Seville, 41092 Seville, Spain; fperdigones@us.es

Keywords: lab-on-PCB; printed circuit board; biomedical applications; electronics; sensors; biosensors; actuators; lab-on-chip; microfluidics

The use of Printed Circuit Boards (PCBs) has seen a remarkable growth over the last decade, with applications in engineering, medicine, biology, chemistry, etc. For example, many biosensors can be fabricated using Printed Circuit Board substrates, resulting in lab-on-PCB devices. In addition, these PCB-based biosensors have been integrated into microfluidic platforms to develop Lab-on-PCB systems [1]. The use of PCBs for developing lab-on-chip devices was proposed in 1996 [2], and the term "Lab-on-PCB" was coined in October 2014 [3]. Apart from sensors and biosensors, the lab-on-PCB devices can include actuators, such as microheaters [4], and devices for microfluid handling, for example, electrowetting on dielectric devices (EWOD) [5]. Since 1996, many lab-on-PCB devices have been developed. However, the majority of the contributions have been published during the last five years. These new contributions focus on the development of applications of current interest. Therefore, we can consider it as an emerging technology.

The lab-on-PCB devices are an interesting improvement of lab-on-chip devices due to the inexpensive characteristic of Printed Circuit Boards, and the possibility of including microfluidic circuits, sensors, and actuators, together with electronics in the same substrate. In addition, these characteristics lead the mass production of biomedical devices, and thus make these devices a very interesting choice from the market point of view.

This Special Issue comprises 10 original contributions related to several aspects of research and development of lab-on-PCB devices. It includes two review papers and eight regular research papers. One of these paper is related with the integration of new emerging rapid prototyping technique with lab-on-PCB devices. This prototyping technique is vat polymerization with an liquid-crystal-display (LCD) as mask, also named masked stereolithography (mSLA) [6]. The technique is available with good resolutions for microfluidics (down to 35 µm). The work reported in [7] describes a technology which creates microfluidics on a Printed Circuit Board substrate using a mSLA printer. The authors describe the steps of the production process, so that, the procedure can be carried out using commercially available printers and resins. This includes the structuring of the copper layer of the PCB and the fabrication of the microchannels layer on top of the Printed Circuit Board. In this respect, the authors fabricated a conductivity sensor chip where several electrode arrangements were demonstrated. They demonstrated, among others, the principle of salinity measurements using seawater and different electrode arrangements using the fabricated microfluidic chip. This technique is a very interesting choice for low cost and rapid prototyping microfluidic devices with integrated sensors.

An important aspect to develop integrated sensors and actuators on lab-on-PCB devices consists in studying the integration method of components. The study of the physical reliability of electronic components are important, principally for thermal processes. In this respect, the work reported in [8] performs a comparative thermal analysis of the cooling efficiency of a surface mounted device (SMD) resistor and an embedded component on a Printed Circuit Board substrate. The authors created a model of heat distribution on a PCB, which took into account the both conductive and convective heat exchange, and

check for updates

Citation: Perdigones, F. Editorial for the Special Issue on Lab-on-PCB Devices. *Micromachines* **2022**, *13*, 1001. https://doi.org/10.3390/mi13071001

Received: 22 June 2022
Accepted: 24 June 2022
Published: 25 June 2022

they confirmed the behavior with experiments. The results show, among others, that the temperature of the embedded component was less than the temperature of the SMD component under natural convection. These results could be taken into account when integrating sensors, actuators or electronics components on lab-on-PCB devices.

Another interesting issue related to PCB-based devices for biomedical applications is the study of the suitability of commercially available biocompatible materials. The work reported in [9] studied the biocompatibility of a commercial Printed Circuit Board for organotypic retina cultures. The authors cultured retinal explants over the white solder mask of a commercial company, on open and closed systems with promising results, that is, the cell viability data show that the white solder mask had no cytotoxic effect on the culture. These results mean a starting point for fabricating microelectrodes arrays (MEAs) on PCB, for regenerative medicine to develop inexpensive methods for improving vision.

Regarding the array of microelectrodes, the work reported in [10] shows a transition from a static MEA (no flow) to a dynamic MEA (continuous flow), assuring a homogeneous transfer of an electrolyte solution (KCl solution) across the measurement chamber. This microfluidic chamber is designed for ensuring both continuous and uniform flow of medium across the complete surface of the chip. The optimum design includes a rounded-corner microfluidic chamber with three inlets and three outlets following the electrode periphery. The results showed, among others, that the change in the voltage noise was minimal when changing from a static MEA to a dynamic one. This is specially interesting for developing long-term cell and tissue cultures using Printed Circuits Boards as substrate.

One of the most useful devices for biomedical applications are the biosensors, particularly electrochemical biosensors. Therefore, the achievement of a robust development of biosensor using PCB substrates is mandatory for a successful future of lab-on-PCB devices. In this respect, the work reported in [11] provide a very good guide for fabricating PCB-based electrochemical biosensors. The authors analyzed the critical technological considerations, allowing the use of Printed Circuit Boards as reliable electrochemical sensing platforms. Both electrochemical and physical characterization showed that organic and inorganic sensing electrode surface includes contaminants which can be removed using several pre-cleaning techniques. In this respect, the authors proposed pre-treatment rules to fabricate PCB-based electrodes for electrochemical biosensors. They demonstrated the applicability of that methodology both for labelled protein and label-free nucleic acid biomarker quantification.

As previously commented, lab-on-PCB devices have very interesting characteristics for the biomedical applications. In particular, one of the most important application is the DNA amplification [12–14]. The device reported in [15] developed a lab-on-PCB for isothermal recombinase polymerase amplification (RPA) of E. coli gDNA using a commercially available 4-layer Printed Circuit Board as substrate. The device includes an integrated microheater fabricated using the copper layer of the substrate. In addition, the microheater is used as temperature sensor. The device also include a copper plate for temperature uniformity. The authors demonstrated the amplification using electrophoresis. The microfluidic chip is intended to be integrated with biosensors in a PCB substrate for the development of inexpensive point-of-care molecular diagnostics platforms.

The previously commented work performed the electrophoresis using an external device. In this respect, the device reported in [16] allows to perform the electrophoresis process using a lab on PCB. The device includes a conductivity sensor to detect the filling with Tris-acetate-EDTA (TAE) buffer. Once the TAE is detected, the device starts a controlled heating process using an integrated microheated and a SMD resistor. In addition, the device includes an optical control of the agarose preparation to finish the process. In order to do so, the authors used a Light-Dependent-Resistor (LDR) integrated in the system. Finally, the PCB substrate includes two electrodes for performing the electrophoresis. The complete system can be considered semi-automatic. However, several parts are completely automatic. This lab-on-PCB device for electrophoresis is intended to be integrated with a lab-on-PCB thermocycler to develop the complete biological process in a lab-on-PCB system.

Related to PCB-based electrophoresis applications, the device reported in [17] describes a novel planar grounded capacitively coupled contactless conductivity detector (PG-C4D) for microchip electrophoresis. The system was composed of a PCB substrate with electrodes and a poly(methyl methacrylate) (PMMA) microfluidic circuit. The reported device had lower stray capacitance than traditional capacitively coupled contactless conductivity detectors. Therefore, the baseline intensity and noise amplitude of the detection cell were smaller. This characteristic implies a higher detection sensitivity, signal-to-noise ratio, and repeatability. This proposed PCB-based PG-C4D device shows an interesting potential for electrophoresis, including, among others, the detection of industrial wastewater or environmental conditions.

An important topic of the microfluidic devices for biomedical applications, and specially for lab-on-PCB devices is the microfluidic handling of small volumes of fluids [18]. The need of microheating, sensing, and micromixing, in different parts of the PCB-based platforms makes the control of liquids necessary. The Special Issue on lab-on-PCB devices includes a review paper which describes the active fluid manipulation methods for lab-on-PCB devices, mentioning their main characteristics from the market point of view [19]. Among others, the author describes the external impulsion devices (syringe or peristaltic pumps); pressurized chambers for displacement of liquid samples and reagents; EWOD; and electro-osmotic and phase-change-based flow driving. This review is an attractive summary for researchers to chose an appropriate fluid manipulation method if they decide to use Printed Circuit Board substrates.

The potential of the lab-on-PCB devices lies on the characteristics of the different layers which compose the Printed Circuit Board substrates. In this respect, the review paper [20] describes the fabrication opportunities that Printed Circuit Boards offer for electronic and biomedical engineering. The authors comment the alternative uses of copper, gold, silver, and Flame Retardant-4 layers. In addition, they mention the use of vias, solder masks and both rigid and flexible substrates. These characteristics have been used to develop sensors, biosensors and actuators, and specially for PCB-based microfluidic platforms. The development of lab-on-PCB devices can still benefit from these alternative uses of printed circuit board layers, allowing the exploitation of commercial PCB-based biomedical and biochemical platforms.

I would like to thank all the authors for submitting their papers to the Special Issue "Lab on PCB Devices". I also thank all the reviewers and editors for spending their time to improve the quality of the submitted papers.

Funding: This work has been funded by regional government Junta de Andalucía (Consejería de Economía y Conocimiento), Plan Andaluz de Investigación, Desarrollo e Innovación (PAIDI 2020) with the project "Sistema para la amplificación y detección de fragmentos de ADN empleando PCR en Lab-on-chip (PCR-on-a-Chip)", reference project "P18-RT-1745". Universidad de Sevilla.

Conflicts of Interest: The author declare no conflicts of interest.

References

1. Zhao, W.; Tian, S.; Huang, L.; Liu, K.; Dong, L. The review of Lab-on-PCB for biomedical application. *Electrophoresis* **2020**, *41*, 1433–1445. [CrossRef] [PubMed]
2. Lammerink, T.; Spiering, V.; Elwenspoek, M.; Fluitman, J.; Van den Berg, A. Modular concept for fluid handling systems. A demonstrator micro analysis system. In Proceedings of Ninth International Workshop on Micro Electromechanical Systems, San Diego, CA, USA, 11–15 February 1996; pp. 389–394.
3. Aracil, C.; Perdigones, F.; Moreno, J.M.; Luque, A.; Quero, J.M. Portable Lab-on-PCB platform for autonomous micromixing. *Microelectron. Eng.* **2015**, *131*, 13–18. [CrossRef]
4. Haci, D.; Liu, Y.; Nikolic, K.; Demarchi, D.; Constandinou, T.G.; Georgiou, P. Thermally Controlled Lab-on-PCB for Biomedical Applications. In Proceedings of the 2018 IEEE Biomedical Circuits and Systems Conference (BioCAS), Cleveland, OH, USA, 17–19 October 2018; pp. 1–4.
5. Yi, Z.; Feng, H.; Zhou, X.; Shui, L. Design of an open electrowetting on dielectric device based on printed circuit board by using a parafilm m. *Front. Phys.* **2020**, *8*, 193. [CrossRef]

6. Navarrete-Segado, P.; Tourbin, M.; Frances, C.; Grossin, D. Masked stereolithography of hydroxyapatite bioceramic scaffolds: From powder tailoring to evaluation of 3D printed parts properties. *Open Ceram.* **2022**, *9*, 100235. [CrossRef]
7. Gassmann, S.; Jegatheeswaran, S.; Schleifer, T.; Arbabi, H.; Schütte, H. 3D Printed PCB Microfluidics. *Micromachines* **2022**, *13*, 470. [CrossRef] [PubMed]
8. Korobkov, M.; Vasilyev, F.; Mozharov, V. A Comparative Analysis of Printed Circuit Boards with Surface-Mounted and Embedded Components under Natural and Forced Convection. *Micromachines* **2022**, *13*, 634. [CrossRef] [PubMed]
9. Urbano-Gámez, J.D.; Valdés-Sánchez, L.; Aracil, C.; de la Cerda, B.; Perdigones, F.; Plaza Reyes, Á.; Díaz-Corrales, F.J.; Relimpio López, I.; Quero, J.M. Biocompatibility Study of a Commercial Printed Circuit Board for Biomedical Applications: Lab-on-PCB for Organotypic Retina Cultures. *Micromachines* **2021**, *12*, 1469. [CrossRef] [PubMed]
10. Ribeiro, M.; Ali, P.; Metcalfe, B.; Moschou, D.; Rocha, P.R. Microfluidics Integration into Low-Noise Multi-Electrode Arrays. *Micromachines* **2021**, *12*, 727. [CrossRef] [PubMed]
11. Zupančič, U.; Rainbow, J.; Estrela, P.; Moschou, D. Utilising Commercially Fabricated Printed Circuit Boards as an Electrochemical Biosensing Platform. *Micromachines* **2021**, *12*, 793. [CrossRef] [PubMed]
12. Kaprou, G.D.; Papadopoulos, V.; Loukas, C.M.; Kokkoris, G.; Tserepi, A. Towards PCB-based miniaturized thermocyclers for DNA amplification. *Micromachines* **2020**, *11*, 258. [CrossRef] [PubMed]
13. Kaprou, G.D.; Papadopoulos, V.; Papageorgiou, D.P.; Kefala, I.; Papadakis, G.; Gizeli, E.; Chatzandroulis, S.; Kokkoris, G.; Tserepi, A. Ultrafast, low-power, PCB manufacturable, continuous-flow microdevice for DNA amplification. *Anal. Bioanal. Chem.* **2019**, *411*, 5297–5307. [CrossRef] [PubMed]
14. Zhang, C.; Xu, J.; Ma, W.; Zheng, W. PCR microfluidic devices for DNA amplification. *Biotechnol. Adv.* **2006**, *24*, 243–284. [CrossRef] [PubMed]
15. Georgoutsou-Spyridonos, M.; Filippidou, M.; Kaprou, G.D.; Mastellos, D.C.; Chatzandroulis, S.; Tserepi, A. Isothermal Recombinase Polymerase Amplification (RPA) of E. coli gDNA in Commercially Fabricated PCB-Based Microfluidic Platforms. *Micromachines* **2021**, *12*, 1387. [CrossRef] [PubMed]
16. Urbano-Gámez, J.D.; Perdigones, F.; Quero, J.M. Semi-Automatic Lab-on-PCB System for Agarose Gel Preparation and Electrophoresis for Biomedical Applications. *Micromachines* **2021**, *12*, 1071. [CrossRef] [PubMed]
17. Wang, J.; Liu, Y.; He, W.; Chen, Y.; You, H. A Novel Planar Grounded Capacitively Coupled Contactless Conductivity Detector for Microchip Electrophoresis. *Micromachines* **2022**, *13*, 394. [CrossRef] [PubMed]
18. Nguyen, N.T.; Hejazian, M.; Ooi, C.H.; Kashaninejad, N. Recent advances and future perspectives on microfluidic liquid handling. *Micromachines* **2017**, *8*, 186. [CrossRef]
19. Perdigones, F. Lab-on-PCB and Flow Driving: A Critical Review. *Micromachines* **2021**, *12*, 175. [CrossRef] [PubMed]
20. Perdigones, F.; Quero, J.M. Printed Circuit Boards: The Layers' Functions for Electronic and Biomedical Engineering. *Micromachines* **2022**, *13*, 460. [CrossRef] [PubMed]

 micromachines

Review

Printed Circuit Boards: The Layers' Functions for Electronic and Biomedical Engineering

Francisco Perdigones * and **José Manuel Quero**

Electronic Engineering Department, University of Seville, 41092 Sevilla, Spain; quero@us.es
* Correspondence: fperdigones@us.es

Abstract: This paper describes the fabrication opportunities that Printed Circuit Boards (PCBs) offer for electronic and biomedical engineering. Historically, PCB substrates have been used to support the components of the electronic devices, linking them using copper lines, and providing input and output pads to connect the rest of the system. In addition, this kind of substrate is an emerging material for biomedical engineering thanks to its many interesting characteristics, such as its commercial availability at a low cost with very good tolerance and versatility, due to its multilayer characteristics; that is, the possibility of using several metals and substrate layers. The alternative uses of copper, gold, Flame Retardant 4 (FR4) and silver layers, together with the use of vias, solder masks and a rigid and flexible substrate, are noted. Among other uses, these characteristics have been using to develop many sensors, biosensors and actuators, and PCB-based lab-on chips; for example, deoxyribonucleic acid (DNA) amplification devices for Polymerase Chain Reaction (PCR). In addition, several applications of these devices are going to be noted in this paper, and two tables summarizing the layers' functions are included in the discussion: the first one for metallic layers, and the second one for the vias, solder mask, flexible and rigid substrate functions.

Keywords: Printed Circuit Board (PCB); biomedical; electronic; engineering

Citation: Perdigones, F.; Quero, J.M. Printed Circuit Boards: The Layers' Functions for Electronic and Biomedical Engineering. *Micromachines* **2022**, *13*, 460. https://doi.org/10.3390/mi13030460

Academic Editor: Nam-Trung Nguyen

Received: 20 February 2022
Accepted: 14 March 2022
Published: 17 March 2022

1. Introduction

The challenge of interconnection between electrical and electronic components led the development of several methods, which can be used to perform this task in a reliable and inexpensive manner. This took place during the end of the 19th century and at the beginning of the 20th century [1,2]. Previously, the interconnection was performed manually using wires. These methods provide the base and origin of the current Printed Circuit Board (PCB).

It is difficult to establish the beginning and inventor of the first Printed Circuit Board. This is because it has the characteristics of several inventions. The first technique used to perform something similar to a double-side PCB was developed by Baynes in 1888 [3] using exposing and etching techniques. However, the functionality of the result was purely ornamental, far from an electronic interconnection. Several years later, Hanson (1903) [4] developed the patent "Electric cable", where an insulating substrate with integrated wires was intended for electrical purposes. This concept is similar to the current single and multilayer PCBs. After that, in 1913, Arthur Berry presented several patents [5] "Improvements in or relating to Electric Heating Apparatus", where the etching technique is used for electro-thermal applications: for fabricating heaters. These thermal devices could be considered one of the first integrated actuators on an insulating substrate. Then, in 1925, Charles Ducas [6] developed a panel containing the metalized lines and subjected it to electroplating to deposit the additional metal. About ten years later, Paul Eisner [7] defined a method to fabricate a stack of metal and insulating materials, which could be considered the first PCB. Finally, Hutters [8] patented the "electrical component mounting device" using hole technology in 1955, and Gabrick [9] defined a composition of the solder mask.

After these developments, a huge quantity of PCBs were fabricated for many different applications. At present, the majority of devices include an electronic part, and this part is made of Printed Circuit Boards; for example, devices for medical applications, such as defibrillators, anesthesia machines, electrocardiogram machines, electrosurgical units and pacemakers to name a few. The devices for industrial equipment also include PCBs; for example, power supplies, computer numerical control (CNC) machines, power inverters or solar power cogeneration devices, to name a few. In addition, automotive, aerospace/satellital and maritime devices include PCBs as electronic substrates for the circuits; for instance, navigation systems, microcontrollers, conditioning electronic circuits for sensors and actuators, and communication equipment. Finally, the consumer electronics, such as videogames, electrical appliances, personal computers and smart phones, form a representative example.

The use of PCB substrates for these applications is well-known and well-established. This has led the creation of many commercial companies, which offer a rapid and low-cost fabrication of PCBs with very good characteristics. These advantages, that is, their low cost and precise dimensions, together with the characteristics of the PCB, made this substrate an interesting material to perform applications for which the PCB was not intended. Regarding this, microelectromechanical systems (MEMS), which typically and historically used silicon as base material, adopted the Printed Circuit Board as an alternative material due to its advantage of having one or more copper layers to fabricate microchannels, sensors or actuators, and the ability to include microfluidics and electronics in the same substrate at a low cost [10–15]. These devices were named PCB-MEMS, and formed the origin of much more developments thanks to the application of lab-on-a-chip (LoC) devices.

Lab-on-a-chips are biomedical, biochemical or chemical devices, including several laboratory functions in a substrate with the dimensions of a credit card, or even smaller. They could include microheating, micropumping, temperature sensors, biological parameter sensors, micromixing and different detection systems. The first LoC materials were silicon and glass. These materials are expensive when considering the typical dimensions of a lab-on-a-chip. For that reason, these materials are going to be replaced by polymers, such as polydimethylsiloxane (PDMS) and SU-8 for prototyping purposes, and thermoplastics for industrial applications. The use of these polymeric materials implies a lack of electrical connections to link the device with the control system, and to integrate sensors and actuators with ease. In this respect, PCBs provide functionalities to solve these problems. Furthermore, the secondary functions of PCB are interesting when looking at the fabrication and performance of the devices.

This paper describes the functionalities of the Printed Circuit Boards layers for different applications to the typical one. The alternative uses of the copper layers, the solder mask, the vias and the PCB substrate are described. Representative devices fabricated using these alternative technologies are noted, emphasizing biomedical and electronic engineering. These technologies will be analyzed in the Section 6 and a conclusion is given in the Section 7. It is important to highlight that this paper is not a comprehensive review of PCB-based devices, but a description of devices that use PCBs as a structural and functional material. More importantly, this paper shows the possibilities that the different layers of PCB substrates offer for the development of electronic and biomedical devices.

2. Printed Circuit Boards, Parts, Dimensions and Types

This section summarizes the most important structures of the Printed Circuit Boards, along with their characteristics and dimensions. This section is important to understand the different functionalities that PCBs offer to the user. The generic structure of a Printed Circuit Board (PCB) is shown in Figure 1.

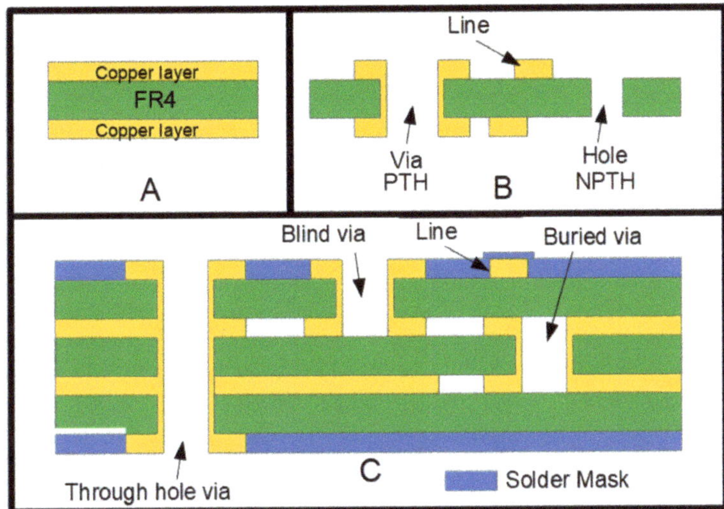

Figure 1. Cross-sectional view of a generic structure of a Printed Circuit Board (PCB). (**A**) Double-side copper layer PCB, where the Flame Retardant 4 (FR4) (green) and the metal (yellow) can be seen. (**B**) Double-side PCB with a copper line, a plated through hole (PTH) via, and a hole (non-plated through hole (NPTH)). (**C**) Four layer PCB with through hole via, blind via, buried via, and a blue solder mask covering the top and bottom layers.

The most common PCB is presented in Figure 1A. As can be seen, it comprises two copper layers (top and bottom) with an intermediate insulation material, which is typically Flame Retardant 4 (FR4). In addition, there is a simpler configuration, with only one copper layer. Figure 1B shows the result of the fabrication process using the double-copper layer, where vias (plated through hole (PTH)) and holes (non-plated through hole (NPTH)) have been fabricated. The vias are characterized as providing electrical and thermal connectivity between the copper layers; in this case, the top and bottom layers.

Figure 1C shows a four-layer PCB. The configuration is similar to the previously noted substrates but includes two additional intermediate copper layers. In this case, the vias can link the four layers, that is, the top, bottom and the intermediate layers. Through vias link the top and bottom layers; blind vias external and intermediate layers; and buried vias link two intermediate layers. The top and bottom surfaces of the final fabricated structure are covered by a blue solder mask. This layer is used to protect the copper lines, releasing only the copper parts used to solder the electronic components. Finally, a silkscreen printing process can be used to name and locate the different components on both the top and bottom surfaces of the PCB. This PCB structure can be used to fabricate more compact electronic circuits. Similarly to this four-layer PCBs, the one- and two-layer PCB can also include solder masks and silkscreen layer.

Apart from these two- and four-layer configurations, the number of total copper layers could be increased. For example, the standard limit that several companies offer is six-layer JLCPCB [16], 14 layers PCBWay [17], e3PCB [18] and PCBgogo [19], and 16 layers Eurocircuits [20] and allPCB [21]. The company Multicircuits boards [22] offers up to eight layers (standard) and up to 28 layers (non standard). Finally, UltimatePCB [23] offers up to 30 copper layers. All these options are for rigid PCB substrates.

In addition, the companies offer the possibility of covering the metal part with tin/lead (hot-air solder leveling (HASL)), gold, silver or Ni/Au/Pd. Regarding the insulating materials, in addition to FR4, different materials can be selected. Furthermore, aluminum is a choice for thermal dissipation applications. Obviously, the aluminum layer is not in contact with the copper layers; a dielectric material is placed between them. Regarding flexible substrates, the materials that could be chosen are polyimide (typical) and polyester (PET) or polyethylene (PE), depending on the company.

Regarding the standard dimensions of the layers, the copper layer can range between 0.5 oz and 13 oz; the insulating layer between 0.17 mm and 7.0 mm; the minimum copper track and spacing 70 µm; minimum diameter vias and hole, 0.15 mm and 0.2 mm, respectively. All these characteristics depend on the manufacturer company. More information can be seen on the manufacturers' website.

These characteristics are used for electronic circuit interconnections. The next sections show alternative functionalities and uses of the different layers of the Printed Circuit Boards.

3. Printed Circuit Boards for Electronics, Sensors and Actuators

As previously noted, the integration of actuators with PCB substrates was performed by Arthur Berry (1913) to fabricate cooking devices. This direct application continues to be interesting, especially for biomedical applications. In addition, thanks to the use of the copper layer, more actuators can be integrated into the PCB, as will be noted below.

3.1. Heaters

The integration of heaters and microheaters is mainly based on the Joule effect. Therefore, the copper layer is patterned to fabricate copper lines for heat dissipation. These heaters were integrated in both rigid [24–28] and flexible [29–31] substrates for deoxyribonucleic acid (DNA) amplification. The temperature should be as uniform as possible in the area of the reaction chamber. To achieve a constant temperature in an area with a low gradient, another copper layer is used as a plate. The microheaters can be fabricated using the top or bottom layer, or even intermediate layers.

For example, the microheaters were used to prepare agarose gel using lab-on-PCB devices [32]. This microheater was fabricated using commercially available PCB substrates, as can be seen in Figure 2. The majority of microheaters are integrated with a thermal sensor to control the temperature set point. In the case of the Figure 2, the sensor is a negative temperature coefficient (NTC) resistor with a surface-mounted device (SMD) package. However, the proper microheater can be used as a temperature sensor [25,28].

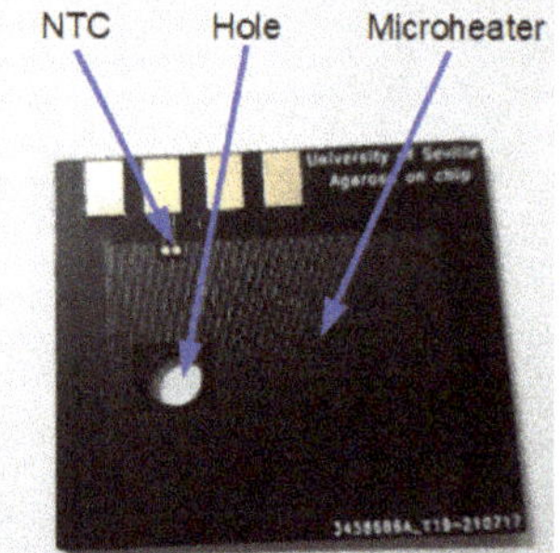

Figure 2. Microheater fabricated using commercially available PCB for agarose gel preparations (reprinted from [32], copyright (2021), Creative Commons License).

In addition, air-flow sensors were developed using PCB-based microheaters [33], as well as devices to study the behaviour of the bubbles [34]. Heaters for the extracellular recording of cardiomyocyte cultures were developed using commercial flexible printed circuit technology [31], and thin copper foil heaters were used to measure the thermal conductivity of polymers [35]. There provide representative examples of microheaters being used for different applications.

3.2. Coils

The coils are used for developing inductors [36]. The most common inductors are intended to be soldered in a PCB, as in Figure 3A. However, this kind of component can be integrated using the copper layers of a Printed Circuit Board.

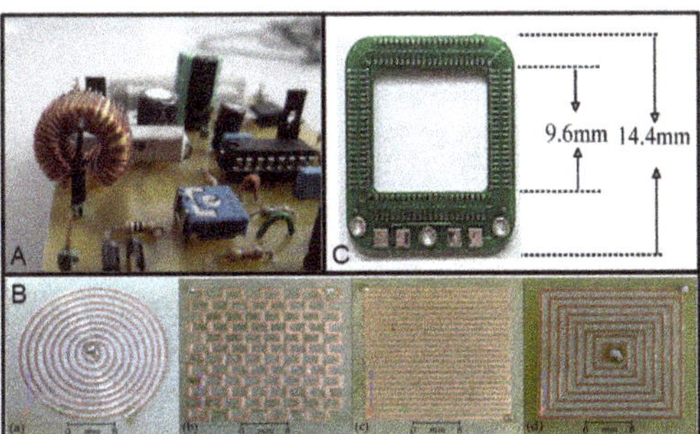

Figure 3. (**A**) Typical inductor assembled to a PCB. (**B**) PCB-based coils with different shapes, (a) Circular planar coil; (b) Mesh planar coil; (c) Meander planar coil and (d) Square planar coil (reprinted from [37], copyright (2018), Creative Commons License). (**C**) Rogowski coil developed on a PCB (reprinted from [38], copyright (2019), Creative Commons License).

The integration of coils on flexible or rigid PCB substrates has been used for wireless power-transmission applications. These coils are fabricated using one copper layer to define the whole structure of the device. The structure is simple: a spiral-shaped copper line in a copper layer, as in Figure 1B, although different topologies are possible [37]. Many devices have been developed using this configuration, for example, printed spiral coils for efficient transcutaneous inductive power transmission [39]. This device is fabricated on a 1-oz copper layer over an FR4 substrate as insulation layer. Similar structures were fabricated for a system with a transmitter and receiver, both of them based on this kind of coil. They are intended for the study of a series of PCB coil matrixes for misalignment-insensitive wireless charging [40]. A current application of these coils is the contactless charger used for handheld devices; for example, smart phones [41]. In addition, the electromagnetic analysis of the alternating current (AC) losses and the practical implementation of PCB planar inductors with a Litz structure were reported [42], as well as the optimization of printed spiral coils for wireless passive sensors [43]. These coils can be fabricated using more than one PCB copper layer; for example, the flow-based electromagnetic-type energy harvester described in [44] included double-sided PCB coils, and the device reported in [45] uses four copper layers two fabricating four coils that are connected in series.

Regarding flexible PCBs, the coils have been used to develop a smartwatch strap wireless power transfer system [46]. To date, the previously noted devices were focused on electronic applications; in contrast, the passive, disposable wireless AC-electroosmotic lab-on-a-PCB, used for particle and fluid manipulation, is an alternative application for biomedical engineering [47].

A different and widely known coil structure is the Rogowski coils. These are electrical devices used to measure the AC current or high-speed current pulses. Typically, they are helicoidal-based metal lines, with a toroid configuration. The conductor cable is encircled by the toroid to measure the current. Many devices were developed using PCB substrates for power electronic applications as current sensors [48–50]. A PCB-base Rogowski coil can be seen in Figure 1C [38]. This kind of coil requires more than one copper layer; for example, the coil reported in [51] is fabricated using four layers, as in Figure 4. This last Rogowski coil is used as a sensor for press-pack insulated gate bipolar transistor chips.

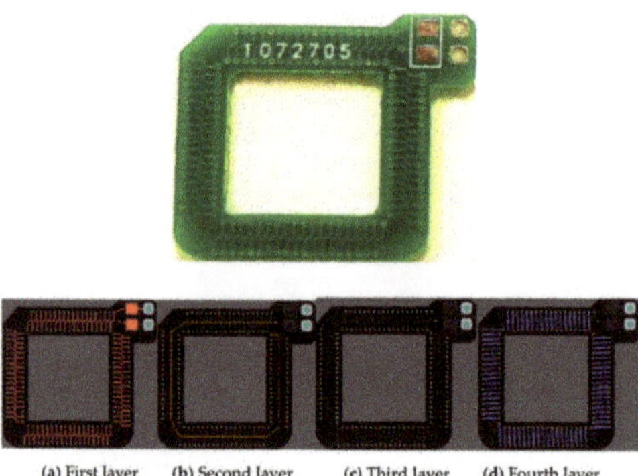

Figure 4. (**Top**) PCB Rogowski coil. (**Bottom**) four-layer board design pattern (reprinted from [51], copyright (2020), Creative Commons License).

A similar geometry to Rogowski coils has been used for fluxgate sensors [52,53]. This low-cost flat fluxgate magnetic field sensor requires a PCB sandwich structure for both single- and double-core versions.

As previously noted, the coils have also been used for biomedical applications; for example, in devices used for malaria detection. The coils were used to enable the translation of magnetic beads, and the agitation and mixing of those beads within a well. The device is composed of six PCB-based coils on the bottom layer and five coils on the top layer [54]. This device is going to be commented on in the biomedical section, but it is interesting to include in this section to provide a general overview of the PCB-based coil for engineering applications.

3.3. *Transformers and Motors*

The contribution of the copper layer of the Printed Circuit Board to the development of a PCB-based transformer and motor is the definition of coils using copper lines, and an FR4 for electrical isolation.

The most common application of the coils is as transformers. The magnetic components have interesting uses in portable electronic applications, for example, as power modules for handheld computers. As the switching frequency of the converter increases, the size of the magnetic core can be reduced. If the switching frequency is high, that is, a few megahertz, the magnetic core can be avoided. Low-cost, coreless, PCB-based transformers for signal and low-power applications have been proposed [55]. Regarding the use of the core, there are other developments [56,57], but additional materials are required, typically ferrite. In addition, these two last developments use a multiple-PCB structure to fabricate the transformers.

The primary and secondary windings are fabricated using the two copper layers of a double-layer PCB [58–63], although there are developments using multiple layers (four layers) PCBs [56,64]. In the first case, the FR4 layer offers an electrical isolation ranging between 15 and 40 kV. A PCB-based transformer integrated with a Printed Circuit Board is shown in Figure 5A; in this transformer, the copper windings are fabricated using external and intermediate copper layers, that is, using a multilayer PCB configuration [65]. The transformer reported in [59] uses self-adhesive ferrite polymer composite sheets to shield the magnetic flux from the transformer windings. The work reported in [60] studies the use of PCB-based transformers with windings on opposite sides to achieve the parasitic inductance cancellation of filter capacitors. Investigations into the use of coreless PCB-based transformers for MOSFET/IGBT gate drive circuits are reported in [64]. The devices are based on a copper coil fabricated using copper layers of the PCB substrate. As can be seen, the contribution of the copper and FR4 layers of PCB—the first one for fabricating the coils and the second one for isolation—make the development of PCB-based transformers possible.

PCB-based motors have attracted increasing interest due to their advantages of a compact size [66], high power density [67], and efficiency [68], design flexibility [69], and low manufacturing costs [70]. Similarly to transformers, PCBs have become a feasible and interesting alternative to conventional round-wire wind. PCB motors can be categorized into two categories depending on the flux direction: radial-flux motors and axial-flux motors [71]. All of them have interesting applications; for instance, PCB-based motors can be used in the development of hard disks [66,68,72]; an axial field permanent magnet motor integrable in the wheel-hub motor of electrical vehicle [73]; for household appliances [70]; for nanosatellites [69] and for a small wind-power system [74].

Multilayer PCBs have been used to fabricate the motors: for example, the PCB stator reported in [67] has 12 layers (Figure 5B); the PCB-based motor for hard disk has six layers in 1-mm-thick PCB, where each layer has nine concentric patterns interconnected by through-holes [68]; a PCB motor intended for use in nanosatellites used a double-layer PCB to integrate the coils [69]; the device reported in [73] requires multilayer PCB (four layers), with 10 PCB-based coils per layer.

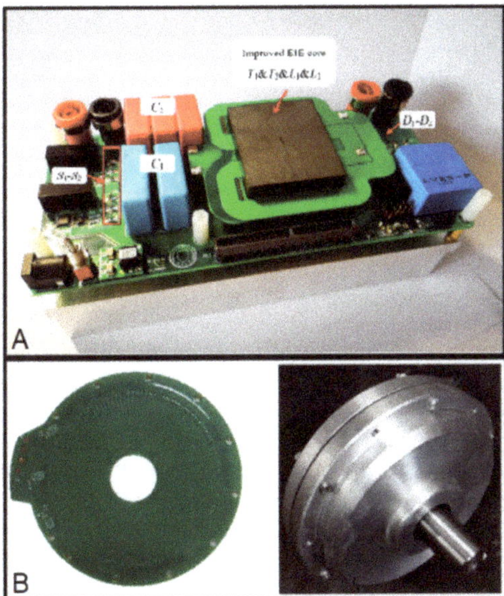

Figure 5. (**A**) PCB-based transformer integrated on a Printed Circuit Board (reprinted from [65], copyright (2020), Creative Commons License). (**B**) Improved PCB stator of a synchronous motor and prototype (reprinted from [67], copyright (2018), Creative Commons License).

Similarly to transformers, the contribution of the copper and FR4 layers of PCB make the development of PCB-based motors possible.

3.4. *Nanogenerator*

The nanogenerators are effective devices for harvesting many kinds of mechanical energy [75,76], especially triboelectric ones, for example, walk energy [77], wave energy [78] or wind power [79], which could be an alternative technology for traditional power generation at large scale. This kind of device can solve the problem of sustainable power and reliable sensing for electronic systems in the Internet-of-Things field. Triboelectric nanogenerators are a promising sustainable power source in self-powered systems.

Printed Circuit Board technology has been used to develop a triboelectric nanogenerator [80–85]. These devices are composed of a stator and a rotor, both fabricated using a single-layer PCB. The stator and the rotor have a circular shape, with the FR4 as the substrate, while radial copper electrodes are fabricated in the copper layer. They are radial-arrayed Cu strips with a unit central angle from 10° [81] to 1° [85], depending on the design of the device. An example of the structure can be seen in Figure 6.

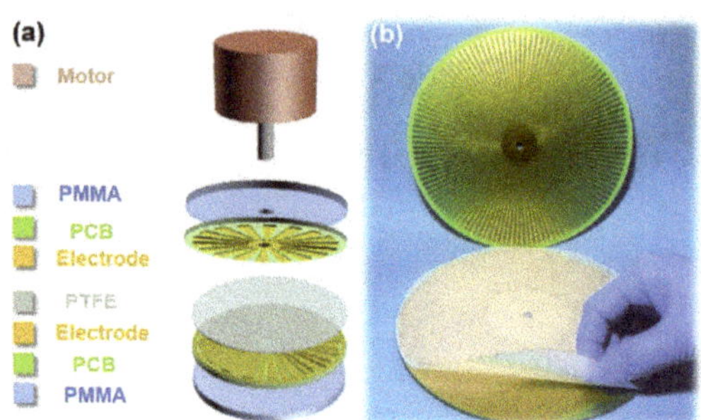

Figure 6. (**a**) Exploded view; (**b**) photograph of a typical rotary disc-shaped triboelectric nanogenerator (reprinted from [83], copyright (2019), with permission from Elsevier).

These devices are intended to be used in intelligent, self-powered, wireless sensing systems [81], among other uses, where the stator copper electrodes are covered with a polytetrafluoroethylene (PTFE) thin film, and Poly(methyl methacrylate) (PMMA) is adhered to the rotor to strengthen the structure. Another example is a self-powered electrospinning system, developed using PCBs [82]. Similarly to the previous one, the stator is fabricated using FR4 and copper, and uses a polymeric film between the stator and rotor. In this case, the rotor is made of kapton and copper, a flexible Printed Circuit Board. PCB-Based triboelectric nanogenerators have also been used to drive self-powered, on-line ion concentration monitors in water transportation [83], Figure 6. This device is composed of a rigid PCB substrate (FR4 and copper) for both the rotor and stator, with a PTFE film between them. It also uses a PMMA sheet to increase the stiffness of the device.

Triboelectric nanogenerators have been developed for sustainable wastewater treatment via a self-powered electrochemical procedure [85]. In this case, the use of the copper layer is similar to the previously noted ones. Finally, a self-powered smart active radio-frequency identification (RFID) tag was integrated with a triboelectric-electromagnetic nanogenerator [84]. It includes a magnet, two PCB plates covered with copper foils, coated with a solder mask acting as a buffer layer, coils with an aluminum-supporting structure, and a polydimethylsiloxane (PDMS) film between the PCB plates. In this case, there was neither a stator nor a rotor; the working principle is based on approaching and separating the two PCB electrodes.

3.5. Fuel Cells

Fuel cells are electrochemical devices, which provide electricity thanks to chemical energy; that is, an electrical current is created using redox reactions. The byproduct of these reactions is water and heat. Several of these devices have been developed, including the proton-exchange membrane fuel cell (PEMFC), alkaline fuel cell (AFC), phosphoric acid fuel cell (PAFC), molten carbonate fuel cell (MCFC), solid oxide fuel cell (SOFC) and direct methanol fuel cell (DMFC) [86]. The main applications of these fuel cells are for power systems [87], cogeneration [88], electric vehicles [89] and portable power systems [90,91].

Many of these devices have been fabricated using PCBs. Both the copper and substrate (rigid and flexible) layers were chosen to built the PCB version of fuel cells. In general, the main function of the copper layers is the development of the anode and cathode current collector; for example, the devices reported in [92–96] were fabricated using rigid PCBs,

Figure 7. The geometry of the anode and cathode openings was studied on [97,98] for PCB devices. In addition, flexible PCBs have been used for fabrication as a current collector [99]. The copper layer corrodes in this kind of device. For that reason, a gold layer covers the copper [100]. This gold layer is an additional material provided by the PCB manufacturer. The gold layers can be seen in Figure 7.

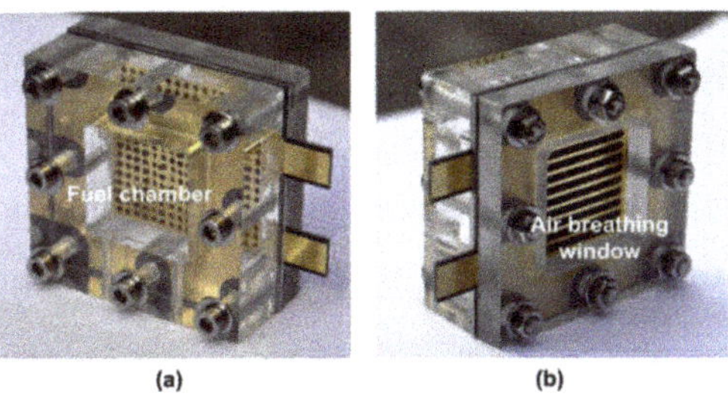

Figure 7. Direct methanol fuel cell: (**a**) fuel chamber with anode; (**b**) air breathing window with the cathode (reprinted from [94], copyright (2015), with permission from Elsevier). As can be seen, the anode and cathode are covered with gold.

Regarding the microfluidic part of the fuel cells, both copper and FR4 layers were used to fabricate the microchannels for the gas [91,92,96,101]. Similarly to other fluidic devices, the vias were used as inlet and outlet ports. Finally, the required electronic traces can be defined in the same PCB [102].

In summary, the copper, gold and FR4 layers make the development of PCB-based fuel cells possible. The copper layer used to fabricate the anode and cathode current collector and microchannels; the gold layer to avoid corrosion, and FR4 to support the structure and fabricating microchannels. In addition, the vias of the PCB can be used as inlet and outlet ports.

3.6. Sensors

A large amount of sensors were developed using Printed Circuit Boards. In this respect, along with the previously commented PCB-based devices, including coils for current and fluxgate sensing, this section describes more functions of the PCB copper layers. In order to do this, the structure of several sensors will be described.

As can be seen, the copper layer of the PCB offers many possibilities to fabricate different devices. This layer continues to be important in the development of sensors.

Many sensors are based on metal electrodes. Therefore, the copper layer of a PCB is a good choice for their fabrication. In this case, a single-layer PCB is enough. An scanning electron microscope (SEM) image of electrodes fabricated on a PCB substrate can be seen in Figure 8A, and the device on Figure 8B.

The electrodes can be used to fabricate capacitive sensors. For example, the work reported in [103] uses the copper layer to develop the electrodes of a tilt sensor where the dielectric material is silicone oil, and the additional structural material is SU-8 photoresist. The two electrodes were fabricated in the same copper layer. Another interesting application of the copper layer in the fabrication of sensors, in this case, a pressure sensor, is the gap definition [104], Figure 8C. This structure uses the copper layer of 18 μm to define the gap between electrodes. Several values of the copper layer can be chosen, if required for different capacitor gaps.

Moreover, the device reported in [105] is a capacitive sensor array for plantar pressure measurement. The device is composed of two rigid PCBs with a double-sided copper layer, with an electromechanical film technology (EMFIT) electroactive ferroelectric film as a dielectric layer. A different pressure sensor was fabricated using both a conductive flexible film and a rigid PCB [106]. In this sensor, one of the electrodes is fabricated using the copper layer, and the second electrode is the conductive flexible film. Finally, pressure sensors were fabricated using liquid crystal polymer with copper (LCP/Cu) Printed Circuit Boards [107]. This sensor includes a 30-mil substrate for the fixed copper electrode, and a flexible 2-mil film for movable electrode. The electrodes can be fabricated on flexible substrates to develop a deformable sensor; for example, the self-powered sensor for human motion detection and gesture recognition [108].

The vias of the PCB were used as capacitive sensor in the reported work on [109,110]. This device is used to detect gas bubbles inside a fluidic flow, where a tube is inserted in the vias, and the fluid flows are inserted inside the tube. Three electrodes were fabricated using the metal of the via.

The temperature sensors were fabricated using PCBs governed by different working principles. For instance, the wireless temperature sensor reported in [111] is fabricated using a double-sided, copper layer PCB, Figure 8D. In this case, the FR4 layer is chosen as material due to its good properties for microwave and RF applications [112]. The working principle is based on two factors: the metal thermal expansion and the dielectric constant of the FR4 depend on the temperature. Similarly to this sensor, the one reported on [113] uses the copper foil on the polyimide (flexible PCB) as a form of thermal resistance to temperature sensing. A different method for sensing temperature consists of using the temperature dependence of a copper line [25]. In this case, the copper line has two functions: as a microheater and the temperature sensor. Finally, the PCB-based device reported in [114] is used as a multisensor platform to measure temperature, conductivity and pressure.

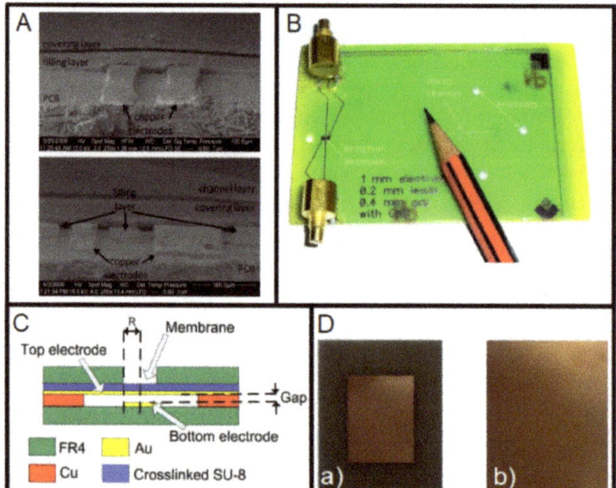

Figure 8. (**A**) Two layers of 30 µm thick dry film photoresist (DFR) laminated on top of electrodes on a PCB (reprinted from [115], copyright (2011), with permission from Elsevier). (**B**) Device with the previously noted electrodes integrated on the PCB (reprinted from [115], copyright (2011), with permission from Elsevier). (**C**) Cross-section view of a pressure sensor with the gap defined using the thickness of the copper layer (copyright (2015) IEEE. Reprinted, with permission, from [104]). (**D**) Sensor fabricated: (a) radiation patch on the upper surface; and (b) metallic ground on the lower surface (reprinted from [111], copyright (2018), Creative Commons License).

Conductivity sensors also require electrodes. The conductivity sensor reported in [116] comprised two planar copper electrodes integrated in a Printed Circuit Board. To determine the conductivity, the sensor measures the resistance between the electrodes when they are submerged in an aqueous solution. PCB-based interdigitated electrodes (IDEs) have also been used to detect icing events [117], to measure the concentrations of nitrate [118] or to measure water content in paper pulp [119], etc.

The copper layer can be processed to achieve interdigitated electrodes. These structures can be used as capacitor, as in the LC displacement sensor reported in [120]. This device integrated both a capacitor and a inductor, both built using the copper layers of the PCB. There are many examples of interdigitated electrodes. They will be detailed in Section 4.2.

Another characteristic of the PCB substrate is its easy integration with surface-mounted devices (SMD); for example, two temperature SMD sensors and a SMD heater were integrated in the PCB to develop a flow sensor [121]. Following this idea, an automotive radar sensor was developed [122]. In this system, the electronic control unit and the RF module are realized in standard multilayer FR4 technology using SMD components.

Printed Circuit Board substrates have also been used to develop accelerometers. The device reported in [123] consists of a metal proof mass, an adhesive tape, and a piece of PCB. The copper layer of the PCB was patterned to fabricate the fixed electrode of the capacitive sensor, and the proof mass was the movable electrode. This device includes the electronic circuit and the sensor in the same PCB substrate. A different device structure was reported in [124], Figure 9. In this case, two rigid Printed Circuit Boards were used to fabricate both the movable and the fixed electrodes. The copper layer of the top movable PCB was used to fabricate the metal plate and the supporting beams. These beams were released by removing the FR4. The FR4over the top metal electrode was not removed in order to define the proof mass and increase the sensitivity. Therefore, the copper layer has two functions: as a metallic electrode and as a movable mechanical structure.

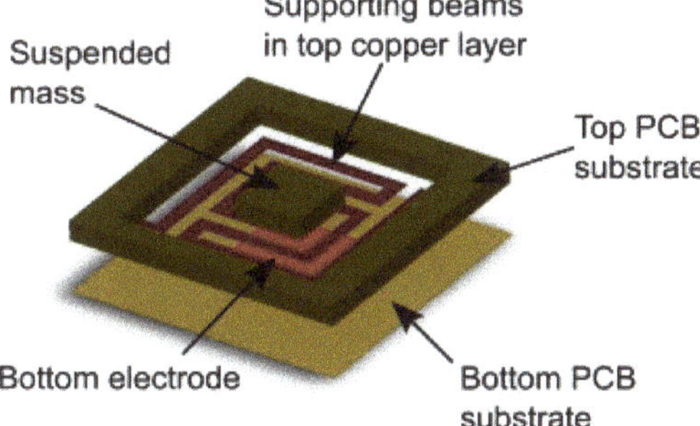

Figure 9. Structure of capacitive PCB-based accelerometer. The beams were fabricated using the copper layer, and the proof mass was defined with the FR4 substrate (reprinted from [124], copyright (2011), with permission from Elsevier).

Finally, a similar device structure was reported on [125], where the fixed electrode was fabricated on a rigid PCB substrate using the copper layer, and the top movable electrode was built on a double-sided, copper-layer, flexible PCB.

The FR4 layer was also used to develop a prototype of a electromagnetic scanning micromirror, integrated with an angle sensor [126], as in Figure 10. The angle sensor was fabricated using one copper layer of the PCB, and the driving coil of the micromirror was fabricated on the opposite copper layer of a double-side PCB. The fabricated device is shown in Figure 11.

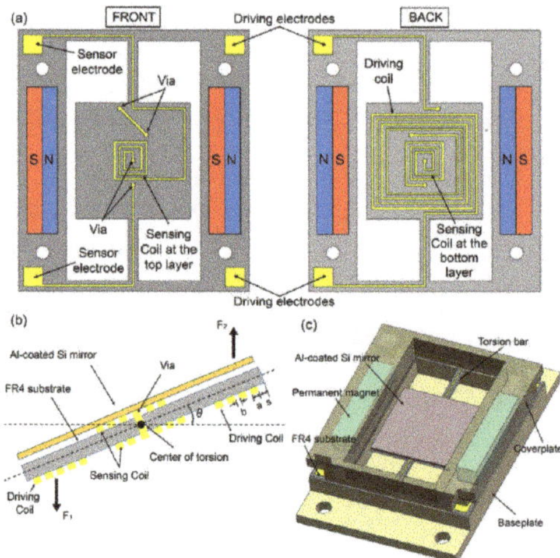

Figure 10. (**a**) Layout of the platform with double-layer copper coils; (**b**) electromagnetic actuation and sensing of the platform with the mirror plate; (**c**) schematic of the assembled scanning micromirror (reprinted from [126], copyright (2018), Creative Commons License).

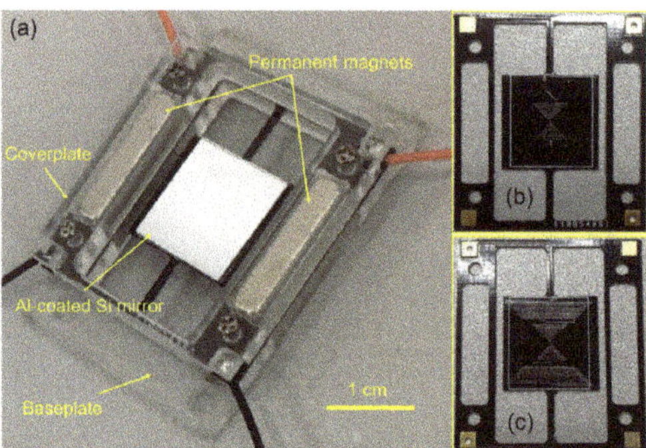

Figure 11. (**a**) Prototype of the electromagnetic scanning micromirror with a plexiglass package; (**b**) front-side view of the platform integrated with copper coils for sensing; (**c**) back-side view of the platform integrated with copper coils for sensing and driving (reprinted from [126], copyright (2018), Creative Commons License).

These devices are examples of sensors where the PCB layers were used to achieve different functions. Moreover, the contribution of the PCB's layers to this kind of devices include cost-effective characteristics, the sensors' easy integration with electronic circuits, and commercially available fabrication.

4. Printed Circuit Boards for Biomedical Applications

In addition to from the typical use of PCBs as an electronic part biomedical devices, the different layers of the PCBs offer interesting possibilities for the development of many different devices. The biosensors, fluid manipulation devices and lab-on-chip have increased their functionalities and commercialization potential thanks to the use of PCBs.

4.1. Fluid Manipulation PCB Devices

Microvalves and micropumps are the most important actuators for fluid manipulation. These devices have been integrated in lab-on-chip devices to control small-volume of fluids, especially for biomedical applications [127–129]. Printed Circuit Boards have been used as a substrate to fabricate LoCs due to their differentiating characteristics [130–133].

The microvalves fabricated using PCBs require an additional material for the microfluidic circuit; for example, SU-8 or PMMA. The microvalves reported in [134–136] used a copper line as a fuse to develop a single-use and normally closed microvalve, Figure 12A. This device was fabricated using SU-8 as an additional material, and PCB as a substrate to fabricate both the microvalve and the electrical connection needed for activation. In contrast, a single-use and normally open microvalve was developed in Reference [137]. In this case, the closing was achieved thermally by the melting of a PMMA microchannel using an integrated microheater. PCB-based heaters have also been used for flow driving [134–136]. In all these cases, the microheater is a copper line, fabricated using the same procedure as the ones commented on Section 3.1. However, in this case, the function of the final device consists of activating the microvalve, that is, the microheater has a secondary function.

There are microvalves based on PCB, which require a different material to perform flow regulation; for example, the PCB-based microvalves reported in Reference [138–141], which use an integrated gold wire as a heater.

The other PCB-based devices for controlling a small volume of fluids are based on electrodes. The fabrication of these electrodes is also based on the typical photolithographic process and etching of the copper layer. Devices for fluid manipulation using electro-osmotic flow [142–144] have been developed, as in Figure 12B. The metal electrodes have been integrated in devices based on electrowetting on a dielectric (EWOD). The structure of an EWOD device is shown in Figure 12C [145]. Some interesting applications of these devices are in pyrosequencing or clinical diagnosis [146,147]. In addition, an electrolytic pump for DNA amplification [148,149] has been fabricated. An example of the PCB-base structure is shown in Figure 12D [150]; as can be seen, it is based on IDEs. Finally, a PCB-based surface acoustic wave (SAW) device for particle manipulation was fabricated using IDEs [151]. This kind of device has an interesting application in cell/droplet manipulation [152]. All of these devices use the copper layers to fabricate the electronic lines. As can be seen, these flow-driving devices are intended for use in biomedical engineering. It is worth highlighting that the device reported in Reference [47] integrated a receiving coil for actuation and an AC electro-osmotic micropump based on IDEs in the same flexible PCB substrate. In addition, the system for malaria detection reported in [54] includes a double-layer PCB, with six coils on the bottom layer and five coils on the top layer.

Similarly to electro-osmotic micropumps, the PCB-based copper lines have been used to fabricate electrodes to move charged particles—in this case, through a polymeric gel. This technique is named electrophoresis. It is used for many biomedical applications; for example, in the separation of DNA fragments [153], for clinical diagnosis [154], for rapid, high-resolution DNA sequencing [155], or for the analysis of drugs in biological fluids [156], etc.

Regarding the PCB-based devices used for electrophoresis, those reported in [32] use copper electrodes plated with gold. Adhesive layers were used to cover the copper electrodes in the device reported in [157]. For capillary electrophoresis, a PCB substrate with platinum wires was used to distribute the electrophoresis voltage [158].

Dielectrophoresis is a similar phenomenon, in which a force is exerted on a dielectric particle subjected to a non-uniform electric field [159]. The fabrication of electrodes using the copper layer of the PCB has made the development of biomedical devices possible, for example, devices for cell manipulation [160], and for the suspension of human tumour cells [161]. In addition, the impact of surface roughness on the dielectrophoretically assisted concentration of microorganisms over PCB-based platforms has been studied [162].

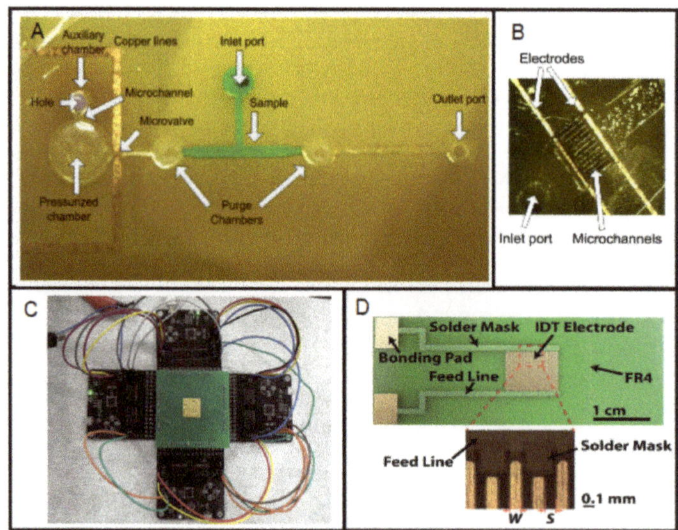

Figure 12. (**A**) Impulsion system based on an SU-8 pressurized chamber and a copper line fuse. (reprinted from [136], copyright (2015), with permission from Elsevier). (**B**) Close view of the electroosmotic part of a PCB-device where the microchannels can be seen (copyright (2013) IEEE. Reprinted, with permission, from [144]). (**C**) Device for fluid manipulation using electrowetting on dielectric on Printed Circuit Board (reprinted from [145], copyright (2020), Creative Commons License). (**D**) Electrochemical PCB-based impulsion chip with detail of the microelectrode fingers (reprinted from [150], copyright (2018), with permission from Elsevier).

Regarding the FR4 layer used for fluid manipulation, the effect of oxygen plasma treatment on the wetting properties of FR4 was investigated [163]. The authors demonstrated that the oxygen plasma treatment of commercially PCB microfluidic structures provides hydrophilic and suitable surfaces for passive microfluidics systems.

Regarding aerodynamic flow control devices, dielectric barrier discharge plasma actuators are typical. This kind of device has been fabricated using Printed Circuit Boards because this decreases the manufacturing cost and weight. Two electrodes are required [164–169]: one of them for high-voltage and second one for the ground. These electrodes are fabricated on opposite sides of the PCB substrate, in which the insulation layer is FR4 and the dielectric breakdown voltage is determined by that material, about 50 kV.

Plasma actuator devices are related to biomedical engineering due to the use of plasma for sterilization, ozonization, surface treatments, or skin treatment by plasma for transdermal drug delivery, etc. [169,170]. In these cases, it is interesting to achieve a jet that is perpendicular to the actuator surface. This is possible with PCB-based dielectric barrier discharge plasma actuators, for example, the device reported in Reference [169].

4.2. Biosensors and Lab on PCB

Biosensors are very important devices for biomedical applications. These devices have been fabricated using PCB substrates, resulting in devices that can be considered Lab-on-PCBs. In addition, PCB-based biosensors can be integrated into a microfluidic platform to develop Lab-on-PCB systems [132]. The use of PCBs for building small laboratories on chip was proposed in 1996 [171], and the term "Lab-on-PCB" was coined in October 2014 [139] (online version). The first device (1996) was a chemical analysis system (μFIA-system). Since 1996, many devices have been developed. However, the majority of the contributions have been published in the last 5 years.

The review of Lab-on-PCB's biomedical applications [132] performed a good analysis of the status of this kind of device regarding its applications, and of the role of the PCB (as a sensor and/or electronic reader). The use of PCB layers in biosensors are going to be presented below.

Generally, the biosensors that use PCBs are the electrochemical ones [172]. These devices are based on electrodes. Thus, the copper layer is used to fabricate them. The copper is not a biocompatible material; therefore, a biocompatible cover is needed to protect the biological samples. In addition, this material has to show a stable voltammogram with a wide potential window. This cover is also provided by the PCB manufacturer; for example, a gold layer deposited by electroplating. The work reported in Reference [172] presented a good study of gold as a material for biochemical electrodes on PCB.

The device reported in Reference [173] is an adhesive and wearable sensor patch used for monitoring sweat electrolytes. It is fabricated using a flexible Printed Circuit Board, where the copper electrodes are covered with Ag/AgCl and paladium. This material (Ag/AgCl) is suitable for developing reference electrodes on PCB [174], as in Figure 13A. This last work demonstrated the good behavior of a pH sensor. In addition, the work reported in Reference [175] also uses silver-based electrodes; in this case, for cancer biomarker detection. Depending on the PCB manufacturer, silver can be ordered as a standard cover. However, the paladium and Ag/AgCl electrodes require additional processes. The HASL cover has also been used to avoid oxidation of the copper, as in the glucose analyzer reported in Reference [176], which is a PCB-based sensor fabricated on a single-side PCB with copper IDEs covered with tin.

The PCB-based chemiresistive carbon dioxide sensor reported in Reference [177] uses silver paste to finish the fabrication of the device. Although this kind of device requires additional processes, it is worth using commercial PCBs to develop them. Many wearable biosensors are based on a flexible printed circuit boards [117,178–181]; for example, the device reported in Reference [179] can be seen in Figure 13B. In addition, a biosensor for SARS-CoV-2 detection was fabricated using flexible PBCs. In that case, graphene was used as an auxiliary material [182].

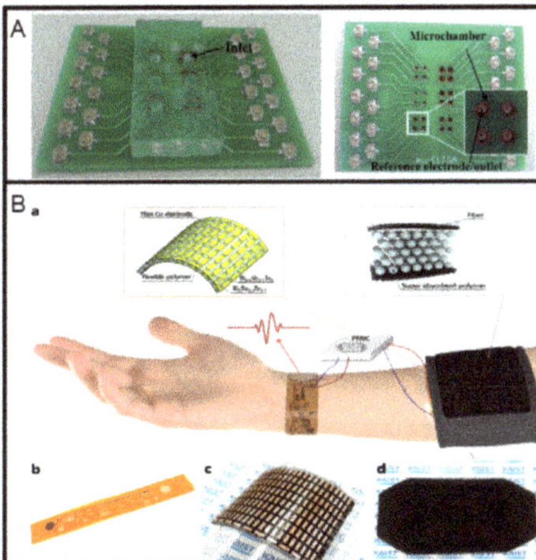

Figure 13. (**A**) Left: Lab-on-PCB integrating microfluidics and PCB microchambers and reference electrodes, right: two-layer PCB before the assembly of microfluidics, comprising microchambers in the top layer and PCB reference electrodes in the bottom layer (reprinted from [174], copyright (2015), Creative Commons License). (**B**) a: schematic diagram of a wearable electrocardiography system, b: flexible electrocardiography module, c: wearable thermoelectric generator, d: polymer-based flexible heat sink (reprinted with permission from [179], copyright (2011), American Chemical Society).

Regarding the use of gold in a biomedical PCB-based sensor, flexible Printed Circuit Board electrodes for the high-resolution mapping of gastrointestinal slow wave activity was reported [183]. This PCB includes gold contacts and copper tracks. The Au-plated electrodes were tested as electrodes (working, counter and reference) [184] to develop a glucose biosensor and exploit the covalent immobilization of commercially fabricated PCB-based electrodes. In addition, the characteristics of the Au-coated test strip for blood glucose measurement using PCB electrodes was reported in Reference [185]. Moreover, the devices described in References [186,187] used both gold and silver PCB-based electrodes. Finally, the PCB-based gold electrodes fabricated using the standard ENIG plating process have been applied to the electrochemical sensing of SARS-CoV-2 amplicons and spike protein [188,189]. As can be seen, the copper, gold and silver layers of the PCB are a good choice for developing biosensors.

The PCB substrate can be used as supporting platform for the system. For instance, the work reported in Reference [190], where an impedimetric transducer is wire-bonded to the PCB and then protected using PDMS, and the SAW viscosity sensor with integrated microfluidics, where the sensor is wire-bonded to the PCB as well [191]. Moreover, the electronic circuit and several sensors can be included in the PCB; for example, the PCB-based system reported in Reference [192] includes a pH indicator, conductivity, sodium, and temperature sensors.

The PCB-based electrodes have also been used to develop conductivity measurements [105,116] for biomedical applications. Among others, the capillary electrophoresis device reported in [115] integrates copper electrodes covered with a dry film photoresist; the copper via holes were used as electrodes on the contactless conductivity sensor for bacterial concentration detection [193], and a similar structure was used for a capillary electrophoresis device [194]. Finally, the most-used electrodes are tIDEs. In addition to the ones presented here to show the different functions of the layers, there are many more examples; for instance, an IDE-based PCB biosensor for the measurement of urea [195], an impedimetric biosensor for the detection of lead (II) based on gold-covered IDEs [196], the IDE-based system for electrochemical sensing of nitrite and taste stimuli reported in Reference [197], where the sensor is fabricated using direct laser writing, which showed a good performance for lower concentrations during taste sensing. The capacitive fringing field sensor for moisture measurement used IDEs covered with a solder mask to avoid contact between the electrodes and the liquid [198]. All these IDEs have straight electrodes. However, different configurations are possible; for example, a circular IDE for cell membrane permeabilization procedures [199].

Lab-on-PCB devices take advantage of every layer in the PCB substrates. Generally, these devices includes sensors, actuators and microfluidics. The microchannels of the microfluidic circuit can be fabricated using the copper lines as walls [10,15]. In addition, in the same work, the authors reported the used of copper lines for the fabrication of microchannels and microheaters; that is, the lines which define the microchannel also define the microheaters. The first lab-on-PCB, including sensors, actuators, electronic components and a microfluidic circuit, was developed for flow injection analysis [11], Figure 14.

Figure 14. The first lab-on-PCB reported by Stefan Gassmann et al. (reprinted from [11], copyright (2007), with permission from Elsevier).

The microchannel can also be defined using the solder mask; for example, the Y-channel reported in [200], the microchannels fabricated in [186,187], and the Lab-on-PCB for the isothermal recombinase polymerase amplification of DNA [25]. This last work included a PCB-based microheater, which simultaneously acts as a temperature sensor. The device is fabricated in a four-layer PCB, where the top and bottom copper layers include the contact pads, and the first and second inner layers define a copper plate for temperature uniformity and the microheater, respectively, Figure 15A . The amplification of DNA has also been performed using several Lab-on-PCB devices. The use of flexible PCB has been studied for both continuous-flow and static-chamber configurations [201]. For example, the continuous-flow PCR microdevices [202–204], and the static-chamber device reported in Reference [24], Figure 15B. All of them include PCB-based microheaters to define a thermal area to perform the PCR. Finally, the work reported in [205] proposed a structure based on two PCBs. The first one was used to define the microchannels on FR4 by milling, and the second one (multilayer PCB) used integrated microheaters, a copper plate for uniformity, a bottom copper layer for wiring, and a top copper layer for electrodes, which were partially defined by the solder mask.

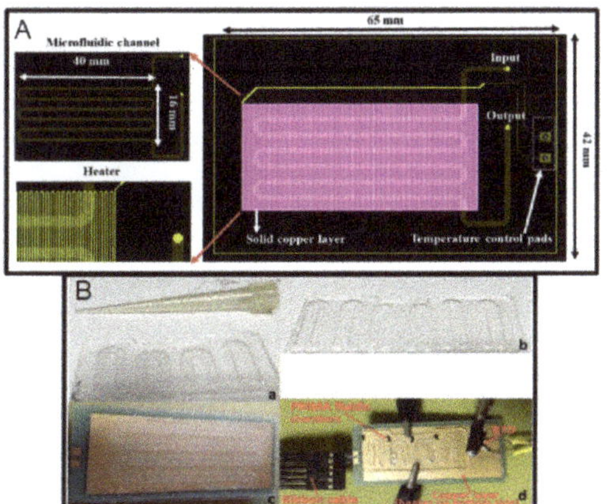

Figure 15. (**A**) Recombinase Polymerase Amplification (RPA)-on-PCB chip design for DNA amplification. The meandering microfluidic channel, the microheater with its electrical pads, and a solid copper layer beneath the microchannel for optimum temperature uniformity are depicted (reprinted from [25], copyright (2021), Creative Commons License). (**B**) a: Poly(methyl methacrylate) PMMA fluidic chip with 4 u-shaped chambers; b: PMMA fluidic with 6 u-shaped chambers; c: PMMA fluidic chip on top of a thin Printed Circuit Board (PCB) microheater with an external temperature-homogenizing copper layer; d: Experimental set-up for temperature measurements during thermocycling of a static micro-polymerase chain reaction (microPCR) chip (reprinted from [24], copyright (2020), Creative Commons License).

Apart from PCR devices, different devices have been developed. Among others, a peptide-nucleic-acid-based Lab-on-PCB diagnostic platform for DNA quantification [206], Figure 16. This device includes a PCB-based sensing layer which consists of two planar, circular gold-plated electrodes and two cylindrical electrodes, used as the fluidic inlet and outlet using vias.

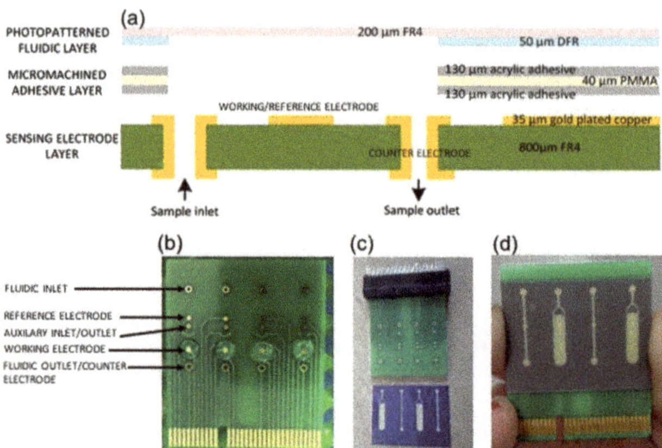

Figure 16. The exploited Lab-on-PCB biosensing platform: (**a**) integrated Lab-on-PCB stack-up; (**b**) Electrochemical Impedance Spectroscopy electrode configuration; (**c**) commercially fabricated PCB biosensing platform; (**d**) sample delivery microfluidics (reprinted from [206], copyright (2019), with permission from Elsevier).

The "flow injection analysis" device reported in Reference [11] was fabricated using a stack of four PCBs. The top PCB included the microchannel, fabricated using the copper lines of the top side. The rest of the PCB substrates, together with a kapton foil, were used to develop a micropump. In addition, the electronic circuit was included in the bottom PCB. Finally, the detection system consisted of a photodiode soldered on the top PCB.

The microfluidic system for the thermal cycling of seawater samples reported in Reference [207] includes an integrated Peltier cell and a SMD temperature sensor for temperature control, and copper areas fabricated using the top copper layer to make the temperature of those areas uniform. In addition, the device includes the the electronic circuit required for the control. The Lab-on-PCB reported in Reference [32] includes a double-layer PCB for agarose gel preparation and electrophoresis. The top copper layer comprises a PCB-base, gold-plated conductivity sensor based on IDEs, and the pads for electronic connections; the bottom layer includes a PCB-based microheater and SMD temperature sensor to control the temperature.

The flow cytometer reported in Reference [208] comprises copper electrodes covered using a cover glass for the detection and enumeration of circulating tumor cells. Similar structures were reported in References [209,210], where the copper electrodes were covered with SU-8, and [211], where the electrodes were covered using PDMS.

The solder masks of Printed Circuit Boards are biocompatible materials for cell and organotypic cultures. For example, the company Multichannel Systems (MultiChannel Systems MCS GmbH, Reutlingen, Germany) offer low-cost, PCB-based microelectrode arrays (MEAs) with one copper layer (Eco MEA), and Elpemer®2467 or PSR-4000 GP01EU as solder masks. The solder mask is used to isolate the metallic lines, releasing the electrodes of the MEA. The biocompatibility of this MEA has been demonstrated for cardiomyocyte cultures, large slices, or whole-heart preparations [212], Figure 17A . In addition, the biocompatibility of the white solder mask PSR-2000 CD02G/CA-25 CD01 has been checked for a retinal continuous culture system, with good results [213]. Finally, the double-side copper layer PCB has also been used to fabricate MEAs; for example, the one provided by Ayanda Biosystems (Ayanda Biosystems SA, Lausanne, Switzerland) [214]. That MEA included a glass with metal electrodes assembled to the PCB, thanks to the bottom copper layer and the PTH vias, as in Figure 17B.

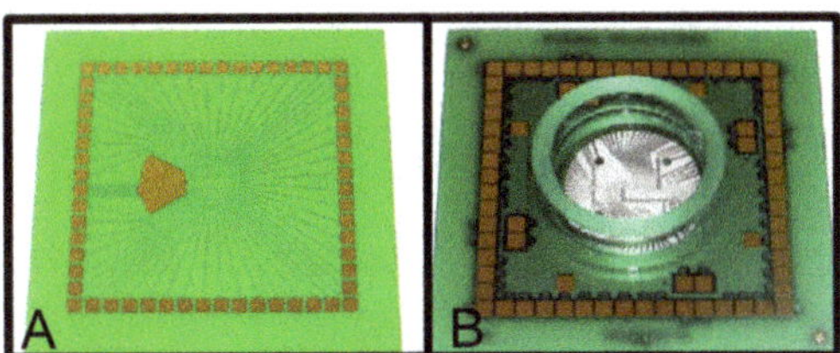

Figure 17. (**A**) Commercial PCB-based microelectrodes arrays of Multichannel Microsystems (model: 60EcoMEA). (**B**) Commercial PCB-based microelectrodes arrays of Ayanda Biosystems™ (model: MEA60 4 × 15 3D).

The use of SMD commercial sensors on the PCB substrates increases the sensitivity of the measurements when their PCB-based counterparts do not provide the required performance. This fact implies an increase in the fabrication complexity, but it is worth if the device fulfills the requirements and works optimally. Among others, the previously noted device includes an SMD temperature sensor [32]; the device reported in reference [215] comprises a silicon-based conductivity sensor and an ion sensitive metal oxide semiconductor (ISFET) pH sensor on the PCB substrate; and the work reported in Reference [190] glued and wire-bonded a transducer array to a Printed Circuit Board.

It is worth mentioning the connection between the portable biomedical devices based on mobile phones or smartphones with Printed Circuit Boards and their layers. For instance, the device used to measure glucose, "HealthPia GlucoPack™ Diabetes Phone", integrated a blood-glucose-monitoring device into the battery pack of a cell phone [216]. In this case, the biosensor is compatible with the Printed Circuit Boards, especially when using the copper layer to fabricate the electrodes. This biosensor is inserted into the mobile to perform the glucose monitoring. Similarly to this device, the one reported in Reference [217] also needs a biosensor that is compatible with commercial PCB fabrication. The device includes the reader for the biosensor and a Bluetooth module for communication with a smartphone. Regarding the wireless communication between the biosensors and smartphones, a system consisting of a smartphone for gas detection [218] was used to describe an example of an adaption of Near-Field-Communication (NFC) technology to a portable and wireless gas-phase chemical sensing. The authors demonstrated the conversion of inexpensive commercial PCB-based NFC tags into chemical sensors. The device reported in Reference [219] is similar. It includes a planar antenna, an SMD NFC microchip and a connector for electrode interface in the same PCB. It was fabricated by a company (PCBWay, Hangzhou, Zhejiang, China) using a rigid FR4-based as substrate, where the copper layer was used to fabricate the passive components. The device is used for both cyclic voltammetry and chronoamperometry. In addition, the device developed in Reference [220] includes commercial NFC tags with resonant circuits consisting of an integrated circuit chip, a chip capacitor, and resistors and an inductance fabricated using the copper layer, on a flexible PCB with polyethylene terephthalate (PET) substrate. The device is intended for biochemical sensing. Finally, the device reported in Reference [221] is similar to the previous one, and includes a PCB-based system for wireless communication and measurement. In this case, the system is an NFC-enabling, smartphone-based portable amperometric immunosensor for hepatitis B virus detection.

As can be seen, electronic components can be fabricated using the copper layer of the PCB, that is, resistors, capacitors and inductances. These components, together with the possibility of soldering chips on the same PCB substrate, lead the development of very interesting applications, some of which are even compatible with smartphones. In addition, the gold, silver or tin/lead layers, and the solder mask provide the solution to avoiding corrosion and oxidation, and to making the biosensor electrodes functional. All the characteristics that are noted in this section open the possibility of developing attractive and complex applications for biomedical applications.

5. Other Uses of PCB Layers

5.1. Antennas

Other interesting uses of the PCB layer include the development of antennas. PCB substrate materials are interesting as a dielectric; for example, the TMM3 and RT/duroid 6002 sheets from Rogers are low-loss materials that provide a good high-frequency performance, with excellent electrical and mechanical properties.

Different types of antennas were fabricated using Printed Circuit Boards. The loop antenna is the simplest one. The PCB-based version is a metallic loop fabricated using one copper layer of the PCB [222]. Regarding patch antennas, several approaches have been reported; for example, the one fabricated only one copper layer [223], where the patch and the ground are fabricated on the same side of the PCB; two copper layers have also been used to fabricate these antennas [224]. In this case, the patch and the ground are defined on opposite copper layers of a double-side PCB. Finally, multilayer PCB was used for developing cavity-backed patch antennas [225], where the top layer includes the patch and one ground plane, the intermediate copper layers include ground planes, and the bottom layer is used to fabricate a microstrip copper line to feed the antenna. The PCBs are used to form a cavity to suppress the surface waves. In addition, they can be used to spread the heat. The slot PCB-based antennas uses two copper layers as well. For instance, the antenna reported in Reference [226] has a metal layer for the microstrip line, and the opposite one for the ground, with an E-shaped slot. Meander antennas were fabricated using PCBs [227]. This device is developed using a double-layer PCB, for the meander and ground, and vias are used to connect them. Finally, the inverted F antennas [228] are based on the use of copper layers with similar functionality to the previously presented antennas.

Figure 18 shows a substrate-integrated, waveguide, cavity-backed slot antenna, fabricated using a PCB substrate. The device includes several metalized vias to avoid energy leakage thanks to the reduction in surface wave propagation. In addition, a pair of triangular-complementary-split-ring slots are included. They are etched on the cavity, which generates a couple of hybrid modes to realize a dual-frequency operation. Although there are no details on the fabrication process, the authors clearly state that they used a single-layered Rogers RT/Duroid 5880 substrate with a thickness of 1.57 mm.

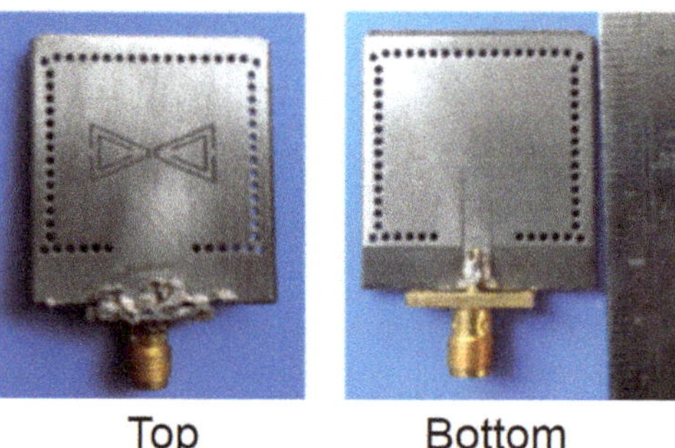

Figure 18. Dual-frequency SIW-based cavity-backed PCB-based antenna. (**Top**): top view where a pair of triangular-complementary-split-ring slots, and vias can be seen; and (**Bottom**): bottom view where the vias can be seen. (reprinted from [229], copyright (2018), with permission from Elsevier).

Wearable antennas are based on flexible Printed Circuit Boards [230]. Since the PCBs allow for the integration of different devices, the biosensor described in Reference [173] offers an interesting functionality—that is, it is an adhesive RFID sensor patch, thanks to the integration of an antenna in the PCB.

Regading RF electronics and antennas, IDE structures for electronic filter banks have been developed using a multiayer PCB to assemble the SAW chip [231]. The authors of this work studied the optimal flip-bonding conditions between the piezoelectric wafer and the PCB.

5.2. Molds

The fabrication of microfluidic devices takes advantage of the use of Printed Circuit Boards. The fabrication of PDMS microfluidic circuit is based on soft lithography; thus, a mold is required. Typically, the molds are fabricated using silicon or SU-8. However, if the dimensional requirements are less demanding, PCB substrates are a good choice. These molds are built using a single-copper-layer PCB [232,233]. In addition, these molds can be used in the hot embossing technique [157,234]. Therefore, thermoplastic materials such as PMMA or polycarbonate can be processed to develop microfluidic devices. Figure 19 shows the photolitographic mask used for fabricating a PCB-based mold, the mold for a serpentine microchannel and the PDMS-fabricated device using that mold.

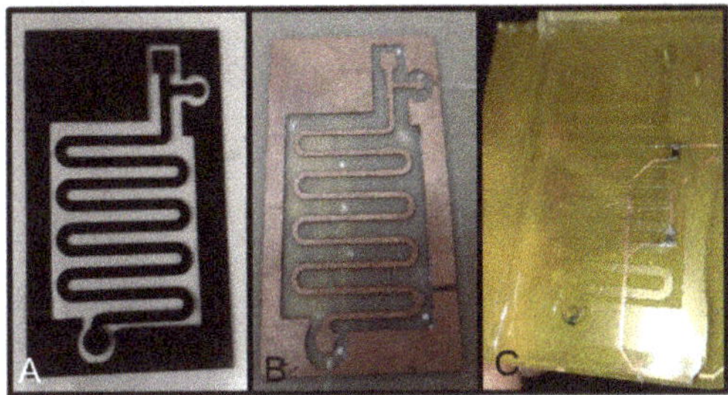

Figure 19. (**A**) Photolitographic mask for fabricating the PCB-based mold. (**B**) Mold for a serpentine microchannel. (**C**) PDMS fabricated device using the mold.

5.3. Flow Focusing

As previously noted, the PCB can be used to fabricate microchannels and chambers. The PCB substrates can be used to fabricate flow-focusing devices. A three-dimensional flow-focusing device for microbubble generation was developed [235], as in Figure 20. This device is fabricated using two single-copper-layer PCBs, where the copper lines are used for the microchannels and microchamber, and the vias are used for inserting the core and shell fluids, that is, gas and water, respectively. In addition, a via is used as a microbubbles outlet.

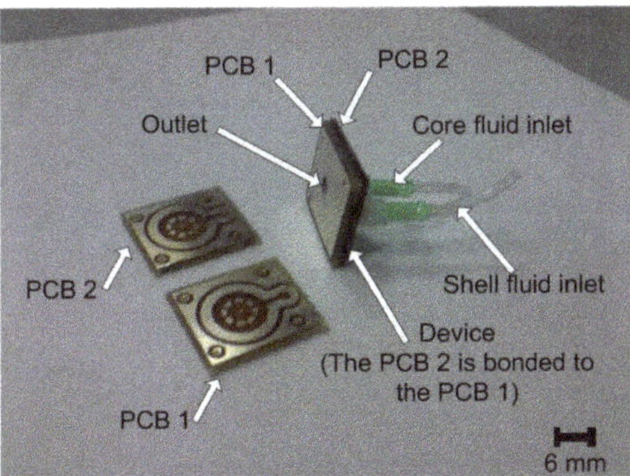

Figure 20. Photograph of the flow-focusing device obtained after the manufacturing process (copyright (2011) IEEE. Reprinted, with permission, from [235]).

The generation of bidimensional particles only requires microchannels. This kind of device has also been fabricated using PCBs [236]. In this case, the copper layer was used to fabricate those microchannels.

5.4. Sacrificial Layer

One of the most important steps in MEMS fabrication is based on the use of a sacrificial layer to fabricate free-standing structures. The copper layer of the PCB can be used as a sacrificial layer to fabricate free-standing SU-8 structures [237]. The chemical etching of the copper does not affect the SU-8. For example, the safety valve reported in Reference [238] was fabricated using the copper as a sacrificial layer; Figure 21A shows the system before the copper etching, and Figure 21B shows the final device. The copper layer thickness defines the gap between the free-standing structure and the substrate. This gap can be selected as a function of the available Cu layer thickness offered by the manufacturer. In addition, the copper layer can be used to release SU-8 structures from the PCB substrate [239,240]. Figure 21C shows a released SU-8 wheel for flow measurement, made of SU-8, by etching a sacrificial copper layer.

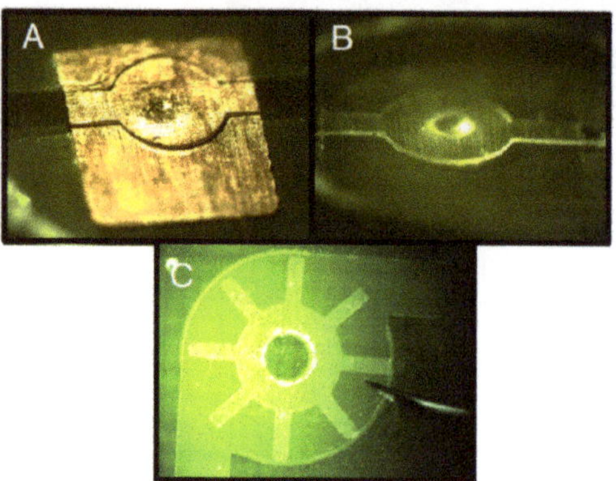

Figure 21. (**A**) Safety valve where the free-standing is not released due to the copper layer (copyright (2010) IEEE. Reprinted, with permission, from [238]). (**B**) Safety valve where the free-standing was released due to the copper layer etching (copyright (2010) IEEE. Reprinted, with permission, from [238]). (**C**) Released wheel for flow measurement made of SU-8 by etching a copper sacrificial layer (copyright (2013) IEEE. Reprinted, with permission, from [240]).

6. Discussion

Thanks to the Printed Circuit Boards, many devices have been fabricated, from purely electronic ones, such as motors or transformers, to biomedical devices, i.e., PCR microdevices. The possibility of integrating different kinds of PCB-based materials led to the fabrication of interesting devices, such as a biosensor with antennas. All these developments are made possible due to the possibilities of all the layers that compose the PCB. The functions of the metallic layers, that is, copper, silver and gold layers, are summarized in Table 1, together with an example. The typical functions of these layers are not included in the Table 1.

The multiple functions of the vias, flexible and rigid substrates are summarized in Table 2, together with an example.

The different metallic layers of the PCBs and their structure make the development of many different devices possible. The layer that offers the most functionalities is the copper layer. The reason for this is that it is a conductive and patternable thin layer. In addition, it is previously deposited when the PCB is bought, with multilayer versions; up to 30 layers are available, depending on the manufacturer. In addition, the vias, the holes and the different materials available for the substrate, mprovide interesting fabrication possibilities. These factors create a technology that can provide devices for different fields, as noted in the paper. It is important to highlight that many devices are fabricated using PCBs. This paper includes representative examples of the devices that take advance of the different fabrication possibilities.

Table 1. Metallic layers functions and devices.

PCB Layer	Function	Device (Example)	Ref.
Gold	Biocompatible electrodes	Glucose biosensor	[184]
	Oxidation prevention	Fuel cell	[94]
Silver	Biocompatible electrodes	Cancer biomarker detection	[175]
Tin/Lead	Oxidation prevention	Glucose sensor	[176]
Copper	Microheater	PCR device	[25]
	Uniform temperature plate	DNA amplification device	[25]
	Microchannels	Microfluidic circuit	[10]
	Microfluidic valve	Impulsion system	[134]
	Capacitive electrodes	Tilt sensor	[103]
	Conductivity electrodes	Conductivity sensor	[116]
	Biochemical electrodes base	pH sensor	[174]
	Anode/cathode current collector	Fuel cell	[94]
	Stator/Rotator	Motor	[67]
	Movable electrodes	Accelerometer	[124]
	EWOD electrodes	Pyrosequencing device	[146]
	Electro-osmotic electrodes	Microfluidic device	[144]
	Electrolytic electrodes	Pumping actuator	[150]
	Gap definition	Pressure sensor	[104]
	Temperature sensor	DNA amplification device	[25]
	Mold	Soft Lithography	[232]
		Hot embossing	[234]
	Sacrificial layer	Safety valve	[238]
	Coils/Spiral	Power transmission	[46]
		Current sensor	[48]
		Fluxgate sensor	[52]
		Antenna	[173]

Table 2. Vias, solder mask, flexible and rigid substrate functions and devices.

PCB Part	Function	Device (Example)	Ref.
Via	Inlet/Outlet fluidic ports	LoP for DNA quantification	[206]
	Capacitive electrode	Bubble detector	[109]
	Conductivity electrode	Bacterial concentration detector	[193]
	Suppress surface waves	Antenna	[225]
	3D flow-focusing outlet	Bubble generator	[235]
Solder mask	Buffer layer	Nanogenerator	[84]
	Microchannel	Microfluidic circuit	[25]
	Non-contact measurements	Moisture sensor	[198]
	Cell culture	MEA	[212]
	Retina cultures	MEA	[213]
Flex substrate	Flexible electrodes	Pressure sensor	[106]
	Wearable devices	Sweat electrolytes sensor	[117]
Rigid	Supporting structure	Conductivity/pH sensor	[215]
	Microchannels (FR4)	PCR device	[205]
	Hydrophilic surface (FR4)	passive microfluidics	[163]
	Proof mass (FR4)	Accelerometer	[124]
	Structural (FR4)	Micromirror	[126]
	Dielectric layer (Rogers)	Antenna	[229]
	Electrical isolation	Transformer	[61]

The opportunity to order PCB for a commercial company facilitates its development by researchers and companies, in the same way that foundries offer their services for silicon and glass fabrication for microelectronics and microsystems. Moreover, PCB processing does not require expensive facilities. This fact allows for the development of devices whose fabrication process does not fit the offered by the manufacturer.

This paper presents the PCB functionalities that have been used to date. However, there are options that have never, to the author's knowledge, been used for a different function than the typical one; for example, aluminum substrate. In this case, the ceramic PCB is used as a supporting structure because it has high mechanical strength and good thermal conductivity [241–244]. Although there companies offering transparent substrates such as glass or PMMA covered with gold, it would be interesting, from the biological perspective, to include PMMA or polycarbonate as a standard material in the typical fabrication process of Printed Circuit Board manufacturing—that is, a PMMA substrate with copper or gold lines, with vias, holes, solder masks and multilayer versions—while maintaining the quality and the cost. This is the reason for not including transparent Printed Circuit Boards. In fact, there are many developments of these devices, as in References [148,245–247].

7. Conclusions

Electronic and electrical engineers have used Printed Circuit Boards since the creation of the substrate. Initially, PCBs were used as connection components, and to connect the substrate to different devices, which, in turn, were composed of Printed Circuit Boards. The multiple fabrication options that PCBs offer makes them useful in the development of electronic components, for example, inductors and capacitors. In addition, the possibility of fabricating electrodes opened up the possibility of developing many different sensors and actuators. Moreover, the ability to include several PCB-based devices, such as antennas and sensors, in the same platform, led to the development of many interesting and complex devices.

Many aspects of the biomedical engineering are closely related with electronic engineering; for example, both of them deal with voltages and amperes, such as intracellular and extracellular potential actions, or the management of current in biosensors, among other applications. In addition, the sensors and actuator provided by electronic engineering make the actuation and control of biomedical devices, such as PCR devices and microdevices, possible.

Printed circuit boards have been used in electronic and biomedical engineering, not only for the development of electronic devices, but also in the fabrication of biosensors, actuators and cell/organotypic culture systems.

New devices and new uses for the different layers of the Printed Circuit Boards are being developed. For example, the discovery of new pathogens and diseases leads research into new devices; for example, the Sars-Cov2 virus pandemic led to the development of PCB devices. Finally, all the possibilities that PCB offers, together with its low cost, will lead to the development of potentially marketable devices, the creation of new companies and an improvement in social welfare in the future.

Author Contributions: Conceptualization, F.P.; methodology, F.P.; validation, F.P.; formal analysis, F.P.; investigation, F.P.; resources, F.P.; writing—original draft preparation, F.P.; writing—review and editing, F.P.; visualization, F.P.; supervision, F.P.; project administration, F.P.; funding acquisition, J.M.Q. and F.P. All authors have read and agreed to the published version of the manuscript.

Funding: This work has been funded by regional government Junta de Andalucía (Consejería de Economía y Conocimiento), Plan Andaluz de Investigación, Desarrollo e Innovación (PAIDI 2020) with the project "Sistema para la amplificación y detección de fragmentos de ADN empleando PCR en Lab-on-chip (PCR-on-a-Chip)", reference project "P18-RT-1745". Universidad de Sevilla.

Institutional Review Board Statement: Not applicable.

Informed Consent Statement: Not applicable.

Data Availability Statement: Data is contained within the article.

Conflicts of Interest: The authors declare no conflict of interest.

References

1. Gilleo, K.; Murray, J. The Definitive History of the Printed Circuit. *PC Fab* **1999**, 1–16.
2. Petherbridge, K.; Evans, P.; Harrison, D. The origins and evolution of the PCB: A review. *Circuit World* **2005** , *31*, 41–45 . [CrossRef]
3. Baynes, J. Method of Etching on One or Both Sides. U.S. Patent 378,423A, 28 February 1888.
4. Hanson, A. Electric Cable. U.S. Patent 782,391A, 14 February 1905.
5. Berry, A. Improvements in or Relating to Electric Heating Apparatus. G.B. Patent 191,314,699A, 25 June 1913.
6. Ducas, C. Electrical Apparatus and Methof of Manufacturing the Same. U.S. Patent 1,563,731A, 2 March 1925.
7. Eisner, P.O.P. Method of Plating Metal Sheets or Plates of Iron or Other Metal. G.B. Patent 431,986A, 18 July 1935.
8. Hutters, E.W. Electrical Component Mounting Device. U.S. Patent 3,002,481, 31 May 1955.
9. Gabrick, A. Solder Mask Composition. U.S. Patent 4,240,945, 23 December 1980.
10. Merkel, T.; Graeber, M.; Pagel, L. A new technology for fluidic microsystems based on PCB technology. *Sens. Actuators A Phys.* **1999**, *77*, 98–105. [CrossRef]
11. Gaßmann, S.; Ibendorf, I.; Pagel, L. Realization of a flow injection analysis in PCB technology. *Sens. Actuators A Phys.* **2007**, *133*, 231–235. [CrossRef]
12. Wego, A.; Pagel, L. A self-filling micropump based on PCB technology. *Sens. Actuators A Phys.* **2001**, *88*, 220–226. [CrossRef]
13. Pagel, L.; Gabmann, S. Microfluidic systems in PCB technology. In Proceedings of the 31st Annual Conference of IEEE Industrial Electronics Society, IECON 2005, Raleigh, NC, USA, 6–10 November 2005; p. 4.
14. Läritz, C.; Pagel, L. A microfluidic pH-regulation system based on printed circuit board technology. *Sens. Actuators A Phys.* **2000**, *84*, 230–235. [CrossRef]
15. Gassmann, S.; Höhne, S.; Pagel, L. Differential pressure flow sensor in PCB-technology. In Proceedings of the 2009 35th Annual Conference of IEEE Industrial Electronics Society, IECON 2009, Porto, Portugal, 3–5 November 2009; pp. 4024–4028.
16. JLCPCB. Available online: https://jlcpcb.com/ (accessed on 8 February 2022).
17. PCBWay. Available online: https://www.pcbway.com/ (accessed on 8 February 2022).
18. e3PCB. Available online: https://www.e3pcb.com/ (accessed on 8 February 2022).
19. PCBgogo. Available online: https://www.pcbgogo.com/ (accessed on 8 February 2022).
20. Eurocircuits. Available online: https://www.eurocircuits.com/ (accessed on 8 February 2022).
21. allPCB. Available online: https://www.allpcb.com/ (accessed on 8 February 2022).
22. Multicircuits. Available online: https://www.multi-circuit-boards.eu/en/index.html (accessed on 8 February 2022).
23. Ultimatepcb. Available online: https://www.ultimatepcb.com/ (accessed on 8 February 2022).
24. Kaprou, G.D.; Papadopoulos, V.; Loukas, C.M.; Kokkoris, G.; Tserepi, A. Towards PCB-based miniaturized thermocyclers for DNA amplification. *Micromachines* **2020**, *11*, 258. [CrossRef]
25. Georgoutsou-Spyridonos, M.; Filippidou, M.; Kaprou, G.D.; Mastellos, D.C.; Chatzandroulis, S.; Tserepi, A. Isothermal Recombinase Polymerase Amplification (RPA) of E. coli gDNA in Commercially Fabricated PCB-Based Microfluidic Platforms. *Micromachines* **2021**, *12*, 1387. [CrossRef]
26. Franco, E.; Salvador, B.; Perdigones, F.; Cabello, M.; Quero, J.M. Fabrication method of lab-on-PCB devices using a microheater with a thermo-mechanical barrier. *Microelectron. Eng.* **2018**, *194*, 31–39. [CrossRef]
27. Tseng, H.Y.; Adamik, V.; Parsons, J.; Lan, S.S.; Malfesi, S.; Lum, J.; Shannon, L.; Gray, B. Development of an electrochemical biosensor array for quantitative polymerase chain reaction utilizing three-metal printed circuit board technology. *Sens. Actuators B Chem.* **2014**, *204*, 459–466. [CrossRef]
28. Tsougeni, K.; Kastania, A.; Kaprou, G.; Eck, M.; Jobst, G.; Petrou, P.; Kakabakos, S.; Mastellos, D.; Gogolides, E.; Tserepi, A. A modular integrated lab-on-a-chip platform for fast and highly efficient sample preparation for foodborne pathogen screening. *Sens. Actuators B Chem.* **2019**, *288*, 171–179. [CrossRef]
29. Lee, D.S.; Choi, O.R.; Seo, Y. A microheater on polyimide substrate for hand-held realtime microfluidic polymerase chain reaction amplification. *Micro Nano Syst. Lett.* **2019**, *7*, 1–4. [CrossRef]
30. Mavraki, E.; Moschou, D.; Kokkoris, G.; Vourdas, N.; Chatzandroulis, S.; Tserepi, A. A continuous flow μPCR device with integrated microheaters on a flexible polyimide substrate. *Procedia Eng.* **2011**, *25*, 1245–1248. [CrossRef]
31. Giovangrandi, L.; Gilchrist, K.H.; Whittington, R.H.; Kovacs, G.T. Low-cost microelectrode array with integrated heater for extracellular recording of cardiomyocyte cultures using commercial flexible printed circuit technology. *Sens. Actuators B Chem.* **2006**, *113*, 545–554. [CrossRef]
32. Urbano-Gámez, J.D.; Perdigones, F.; Quero, J.M. Semi-Automatic Lab-on-PCB System for Agarose Gel Preparation and Electrophoresis for Biomedical Applications. *Micromachines* **2021**, *12*, 1071. [CrossRef]
33. Glatzl, T.; Steiner, H.; Kohl, F.; Sauter, T.; Keplinger, F. Development of an air flow sensor for heating, ventilating, and air conditioning systems based on printed circuit board technology. *Sens. Actuators A Phys.* **2016**, *237*, 1–8. [CrossRef]
34. Tanjung, E.F.; Alunda, B.O.; Lee, Y.J.; Jo, D. Experimental study of bubble behaviors and CHF on printed circuit board (PCB) in saturated pool water at various inclination angles. *Nucl. Eng. Technol.* **2018**, *50*, 1068–1078. [CrossRef]
35. Kazeminejad, H. Thin copper foil heater for measuring the thermal conductivity of polymers. *J. Appl. Polym. Sci.* **2003**, *88*, 2823–2827. [CrossRef]
36. Dezuari, O.; Gilbert, S.; Belloy, E.; Gijs, M. Development of a novel printed circuit board technology for inductive device applications. *Sens. Actuators A Phys.* **1999**, *76*, 349–355. [CrossRef]

37. Gibbs, R.; Moreton, G.; Meydan, T.; Williams, P. Comparison between modelled and measured magnetic field scans of different planar coil topologies for stress sensor applications. *Sensors* **2018**, *18*, 931. [CrossRef]
38. Jiao, C.; Zhang, J.; Zhao, Z.; Zhang, Z.; Fan, Y. Research on small square pcb rogowski coil measuring transient current in the power electronics devices. *Sensors* **2019**, *19*, 4176. [CrossRef]
39. Jow, U.M.; Ghovanloo, M. Design and optimization of printed spiral coils for efficient transcutaneous inductive power transmission. *IEEE Trans. Biomed. Circuits Syst.* **2007**, *1*, 193–202. [CrossRef]
40. Li, J.; Wang, L.; Yin, F. Study on Series Printed-Circuit-Board Coil Matrix for Misalignment-Insensitive Wireless Charging. In Proceedings of the 2019 IEEE Wireless Power Transfer Conference (WPTC 2019), London, UK, 18–21 June 2019; pp. 98–101.
41. Hu, C.H.; Chen, C.M.; Shiao, Y.S.; Chan, T.J.; Chen, T.R. Development of a universal contactless charger for handheld devices. In Proceedings of the 2008 IEEE International Symposium on Industrial Electronics (ISIE 2008), Cambridge, UK, 30 June–2 July 2008; pp. 99–104.
42. Lope, I.; Carretero, C.; Acero, J.; Alonso, R.; Burdio, J.M. Frequency-dependent resistance of planar coils in printed circuit board with litz structure. *IEEE Trans. Magn.* **2014**, *50*, 1–9. [CrossRef]
43. Noroozi, B.; Morshed, B.I. Design and optimization of printed spiral coils for wireless passive sensors. *IET Wirel. Sens. Syst.* **2021**, *11*, 169–178. [CrossRef]
44. Ahmad, I.; Ur Rehman, M.M.; Khan, M.; Abbas, A.; Ishfaq, S.; Malik, S. Flow-based electromagnetic-type energy harvester using microplanar coil for IoT sensors application. *Int. J. Energy Res.* **2019**, *43*, 5384–5391. [CrossRef]
45. Li, Z.; He, X.; Shu, Z. Design of coils on printed circuit board for inductive power transfer system. *IET Power Electron.* **2018**, *11*, 2515–2522. [CrossRef]
46. Jeong, S.; Kim, D.H.; Song, J.; Kim, H.; Lee, S.; Song, C.; Lee, J.; Song, J.; Kim, J. Smartwatch strap wireless power transfer system with flexible PCB coil and shielding material. *IEEE Trans. Ind. Electron.* **2018**, *66*, 4054–4064. [CrossRef]
47. Mirzajani, H.; Cheng, C.; Wu, J.; Ivanoff, C.S.; Aghdam, E.N.; Ghavifekr, H.B. Design and characterization of a passive, disposable wireless AC-electroosmotic lab-on-a-film for particle and fluid manipulation. *Sens. Actuators B Chem.* **2016**, *235*, 330–342. [CrossRef]
48. Shi, Y.; Xin, Z.; Loh, P.C.; Blaabjerg, F. A Review of Traditional Helical to Recent Miniaturized Printed Circuit Board Rogowski Coils for Power-Electronic Applications. *IEEE Trans. Power Electron.* **2020**, *35*, 12207–12222. [CrossRef]
49. Qing, C.; Li, H.-B.; Zhang, M.-M.; Liu, Y.-B. Design and characteristics of two Rogowski coils based on printed circuit board. *IEEE Trans. Instrum. Meas.* **2006**, *55*, 939–943. [CrossRef]
50. Li, Y.; Guo, Y.; Long, Y.; Yao, C.; Mi, Y.; Wu, J. Novel lightning current sensor based on Printed Circuit Board Rogowski coil. In Proceedings of the 2012 International Conference on High Voltage Engineering and Application, (ICHVE 2012), Shanghai, China, 17–20 September 2012; pp. 334–338.
51. Jiao, C.; Zhang, Z.; Zhao, Z.; Zhang, X. Integrated Rogowski Coil Sensor for Press-Pack Insulated Gate Bipolar Transistor Chips. *Sensors* **2020**, *20*, 4080. [CrossRef]
52. Kubík, J.; Pavel, L.; Ripka, P.; Kaspar, P. Low-power printed circuit board fluxgate sensor. *IEEE Sens. J.* **2007**, *7*, 179–183. [CrossRef]
53. Kubik, J.; Pavel, L.; Ripka, P. PCB racetrack fluxgate sensor with improved temperature stability. *Sens. Actuators A Phys.* **2006**, *130*, 184–188. [CrossRef]
54. Kim, C.; Hoffmann, G.; Searson, P.C. Integrated magnetic bead–quantum dot immunoassay for malaria detection. *ACS Sens.* **2017**, *2*, 766–772. [CrossRef]
55. Tang, S.; Hui, S.; Chung, H.H. Evaluation of the shielding effects on printed-circuit-board transformers using ferrite plates and copper sheets. *IEEE Trans. Power Electron.* **2002**, *17*, 1080–1088. [CrossRef]
56. Ho, Y.; Cheng, K.; Kan, K.L.J.; Chung, H.; Lam, W.Y. Planar Printed-Circuit-Board (PCB) Transformers with Active Clamp Flyback Converter for Low Power AC-DC Adapter Application. In Proceedings of the 2020 8th International Conference on Power Electronics Systems and Applications (PESA), Hong Kong, China, 7–10 December 2020; pp. 1–5.
57. Tria, L.A.R.; Alam, K.S.; Zhang, D.; Fletcher, J.E. Comparative study of multicore planar transformers on printed circuit boards. *IET Power Electron.* **2017**, *10*, 1452–1460. [CrossRef]
58. Tang, S.; Hui, S.; Chung, H.S.H. Coreless planar printed-circuit-board (PCB) transformers-a fundamental concept for signal and energy transfer. *IEEE Trans. Power Electron.* **2000**, *15*, 931–941. [CrossRef]
59. Tang, S.; Hui, S.R.; Chung, H.H. A low-profile power converter using printed-circuit board (PCB) power transformer with ferrite polymer composite. *IEEE Trans. Power Electron.* **2001**, *16*, 493–498. [CrossRef]
60. Neugebauer, T.C.; Perreault, D.J. Filters with inductance cancellation using printed circuit board transformers. *IEEE Trans. Power Electron.* **2004**, *19*, 591–602. [CrossRef]
61. Hui, S.; Chung, H.S.H.; Tang, S. Coreless printed circuit board (PCB) transformers for power MOSFET/IGBT gate drive circuits. *IEEE Trans. Power Electron.* **1999**, *14*, 422–430. [CrossRef]
62. Tang, S.; Hui, S.; Chung, H.H. Characterization of coreless printed circuit board (PCB) transformers. *IEEE Trans. Power Electron.* **2000**, *15*, 1275–1282. [CrossRef]
63. Marinova, I.; Midorikawa, Y.; Hayano, S.; Saito, Y. Thin film transformer and its analysis by integral equation method. *IEEE Trans. Magn.* **1995**, *31*, 2432–2437. [CrossRef]

64. Majid, A.; Kotte, H.B.; Saleem, J.; Ambatipudi, R.; Haller, S.; Bertilsson, K. High frequency half-bridge converter using multilayered coreless printed circuit board step-down power transformer. In Proceedings of the 8th International Conference on Power Electronics & ECCE Asia, Jeju, Korea, 29 May–2 June 2011; pp. 1177–1181.
65. Liu, R.; Wang, Y.; Chen, Q.; Han, F.; Meng, Z. Entire magnetic integration method of multi-transformers and resonant inductors for cltlc resonant converter. *Electronics* **2020**, *9*, 1386. [CrossRef]
66. Tsai, M.C.; Hsu, L.Y. Design of a miniature axial-flux spindle motor with rhomboidal PCB winding. *IEEE Trans. Magn.* **2006**, *42*, 3488–3490. [CrossRef]
67. Wang, X.; Lu, H.; Li, X. Winding design and analysis for a disc-type permanent-magnet synchronous motor with a PCB stator. *Energies* **2018**, *11*, 3383. [CrossRef]
68. Jang, G.; Chang, J. Development of an axial-gap spindle motor for computer hard disk drives using PCB winding and dual air gaps. *IEEE Trans. Magn.* **2002**, *38*, 3297–3299. [CrossRef]
69. Wu, J. Design of a miniature axial flux flywheel motor with PCB winding for nanosatellites. In Proceedings of the 2012 International Conference on Optoelectronics and Microelectronics, (ICOM 2012), Changchun, China, 23–25 August 2012; pp. 544–548.
70. Marignetti, F.; Volpe, G.; Mirimani, S.M.; Cecati, C. Electromagnetic design and modeling of a two-phase axial-flux printed circuit board motor. *IEEE Trans. Ind. Electron.* **2017**, *65*, 67–76. [CrossRef]
71. Taqavi, O.; Mirimani, S.M. Design aspects, winding arrangements and applications of printed circuit board motors: A comprehensive review. *IET Electr. Power Appl.* **2020**, *14*, 1505–1518. [CrossRef]
72. Jang, G.; Chang, J. Development of dual air gap printed coil BLDC motor. *IEEE Trans. Magn.* **1999**, *35*, 1789–1792. [CrossRef]
73. Wang, X.; Li, C.; Lou, F. Geometry optimize of printed circuit board stator winding in coreless axial field permanent magnet motor. In Proceedings of the 2016 IEEE Vehicle Power and Propulsion Conference (VPPC), Hangzhou, China, 17–20 October 2016; pp. 1–6.
74. Wang, X.; Pang, W.; Gao, P.; Zhao, X. Electromagnetic design and analysis of axial flux permanent magnet generator with unequal-width PCB winding. *IEEE Access* **2019**, *7*, 164696–164707. [CrossRef]
75. Wang, Z.L.; Lin, L.; Chen, J.; Niu, S.; Zi, Y. *Triboelectric Nanogenerators*; Springer Nature Singapore Pte Ltd.: Singapore, 2016.
76. Wang, Z.L. Triboelectric nanogenerators as new energy technology for self-powered systems and as active mechanical and chemical sensors. *ACS Nano* **2013**, *7*, 9533–9557. [CrossRef] [PubMed]
77. Hou, T.C.; Yang, Y.; Zhang, H.; Chen, J.; Chen, L.J.; Wang, Z.L. Triboelectric nanogenerator built inside shoe insole for harvesting walking energy. *Nano Energy* **2013**, *2*, 856–862. [CrossRef]
78. Hu, Y.; Yang, J.; Jing, Q.; Niu, S.; Wu, W.; Wang, Z.L. Triboelectric nanogenerator built on suspended 3D spiral structure as vibration and positioning sensor and wave energy harvester. *ACS Nano* **2013**, *7*, 10424–10432. [CrossRef] [PubMed]
79. Xie, Y.; Wang, S.; Lin, L.; Jing, Q.; Lin, Z.H.; Niu, S.; Wu, Z.; Wang, Z.L. Rotary triboelectric nanogenerator based on a hybridized mechanism for harvesting wind energy. *ACS Nano* **2013**, *7*, 7119–7125. [CrossRef]
80. Han, C.; Zhang, C.; Tang, W.; Li, X.; Wang, Z.L. High power triboelectric nanogenerator based on printed circuit board (PCB) technology. *Nano Res.* **2015**, *8*, 722–730. [CrossRef]
81. Lu, S.; Gao, L.; Chen, X.; Tong, D.; Lei, W.; Yuan, P.; Mu, X.; Yu, H. Simultaneous energy harvesting and signal sensing from a single triboelectric nanogenerator for intelligent self-powered wireless sensing systems. *Nano Energy* **2020**, *75*, 104813. [CrossRef]
82. Li, C.; Yin, Y.; Wang, B.; Zhou, T.; Wang, J.; Luo, J.; Tang, W.; Cao, R.; Yuan, Z.; Li, N.; et al. Self-powered electrospinning system driven by a triboelectric nanogenerator. *ACS Nano* **2017**, *11*, 10439–10445. [CrossRef]
83. Chen, C.; Wen, Z.; Wei, A.; Xie, X.; Zhai, N.; Wei, X.; Peng, M.; Liu, Y.; Sun, X.; Yeow, J.T. Self-powered on-line ion concentration monitor in water transportation driven by triboelectric nanogenerator. *Nano Energy* **2019**, *62*, 442–448. [CrossRef]
84. Chen, Y.L.; Liu, D.; Wang, S.; Li, Y.F.; Zhang, X.S. Self-powered smart active RFID tag integrated with wearable hybrid nanogenerator. *Nano Energy* **2019**, *64*, 103911. [CrossRef]
85. Chen, S.; Wang, N.; Ma, L.; Li, T.; Willander, M.; Jie, Y.; Cao, X.; Wang, Z.L. Triboelectric nanogenerator for sustainable wastewater treatment via a self-powered electrochemical process. *Adv. Energy Mater.* **2016**, *6*, 1501778. [CrossRef]
86. Kirubakaran, A.; Jain, S.; Nema, R. A review on fuel cell technologies and power electronic interface. *Renew. Sustain. Energy Rev.* **2009**, *13*, 2430–2440. [CrossRef]
87. Lai, J.S.; Ellis, M.W. Fuel cell power systems and applications. *Proc. IEEE* **2017**, *105*, 2166–2190. [CrossRef]
88. Lavernia, A.; Dover, T.; Samuelsen, S. Operational and economic performance analysis of a high-temperature fuel cell cogeneration plant. *J. Power Sources* **2022**, *520*, 230798. [CrossRef]
89. Tanç, B.; Arat, H.T.; Baltacıoğlu, E.; Aydın, K. Overview of the next quarter century vision of hydrogen fuel cell electric vehicles. *Int. J. Hydrogen Energy* **2019**, *44*, 10120–10128. [CrossRef]
90. Wilberforce, T.; Alaswad, A.; Palumbo, A.; Dassisti, M.; Olabi, A.G. Advances in stationary and portable fuel cell applications. *Int. J. Hydrogen Energy* **2016**, *41*, 16509–16522. [CrossRef]
91. O'Hayre, R.; Braithwaite, D.; Hermann, W.; Lee, S.J.; Fabian, T.; Cha, S.W.; Saito, Y.; Prinz, F.B. Development of portable fuel cell arrays with printed-circuit technology. *J. Power Sources* **2003**, *124*, 459–472. [CrossRef]
92. Cleghorn, S.; Derouin, C.; Wilson, M.; Gottesfeld, S. A printed circuit board approach to measuring current distribution in a fuel cell. *J. Appl. Electrochem.* **1998**, *28*, 663–672. [CrossRef]
93. Schmitz, A.; Wagner, S.; Hahn, R.; Uzun, H.; Hebling, C. Stability of planar PEMFC in printed circuit board technology. *J. Power Sources* **2004**, *127*, 197–205. [CrossRef]

94. Yuan, W.; Zhang, X.; Zhang, S.; Hu, J.; Li, Z.; Tang, Y. Lightweight current collector based on printed-circuit-board technology and its structural effects on the passive air-breathing direct methanol fuel cell. *Renew. Energy* **2015**, *81*, 664–670. [CrossRef]
95. Hong, P.; Liao, S.; Zeng, J.; Huang, X. Design, fabrication and performance evaluation of a miniature air breathing direct formic acid fuel cell based on printed circuit board technology. *J. Power Sources* **2010**, *195*, 7332–7337. [CrossRef]
96. Jafri, R.I.; Ramaprabhu, S. Multi walled carbon nanotubes based micro direct ethanol fuel cell using printed circuit board technology. *Int. J. Hydrogen Energy* **2010**, *35*, 1339–1346. [CrossRef]
97. Lim, S.W.; Kim, S.W.; Kim, H.J.; Ahn, J.E.; Han, H.S.; Shul, Y.G. Effect of operation parameters on performance of micro direct methanol fuel cell fabricated on printed circuit board. *J. Power Sources* **2006**, *161*, 27–33. [CrossRef]
98. Obeisun, O.A.; Meyer, Q.; Robinson, J.; Gibbs, C.W.; Kucernak, A.R.; Shearing, P.R.; Brett, D.J. Development of open-cathode polymer electrolyte fuel cells using printed circuit board flow-field plates: Flow geometry characterisation. *Int. J. Hydrogen Energy* **2014**, *39*, 18326–18336. [CrossRef]
99. Kim, S.H.; Cha, H.Y.; Miesse, C.M.; Jang, J.H.; Oh, Y.S.; Cha, S.W. Air-breathing miniature planar stack using the flexible printed circuit board as a current collector. *Int. J. Hydrogen Energy* **2009**, *34*, 459–466. [CrossRef]
100. Guo, J.W.; Xie, X.F.; Wang, J.H.; Shang, Y.M. Effect of current collector corrosion made from printed circuit board (PCB) on the degradation of self-breathing direct methanol fuel cell stack. *Electrochim. Acta* **2008**, *53*, 3056–3064. [CrossRef]
101. Bethapudi, V.; Hack, J.; Trogadas, P.; Cho, J.; Rasha, L.; Hinds, G.; Shearing, P.; Brett, D.; Coppens, M.O. A lung-inspired printed circuit board polymer electrolyte fuel cell. *Energy Convers. Manag.* **2019**, *202*, 112198. [CrossRef]
102. Schulze, M.; Gülzow, E.; Knöri, T.; Reissner, R. Segmented cells as tool for development of fuel cells and error prevention/prediagnostic in fuel cell stacks. *J. Power Sources* **2007**, *173*, 19–27. [CrossRef]
103. Salvador, B.; Luque, A.; Quero, J.M. Microfluidic capacitive tilt sensor using PCB-MEMS. In Proceedings of the 2015 IEEE International Conference on Industrial Technology (ICIT 2015), Seville, Spain, 17–19 March 2015; pp. 3356–3360.
104. Souilah, M.; Chaabi, A.; Perdigones, F.; Quero, J.M.; Flores, G.; Lain, M.R. Fabrication process for PCBMEMS capacitive pressure sensors using the cu layer to define the gap. *IEEE Sens. J.* **2015**, *16*, 1151–1157. [CrossRef]
105. Aqueveque, P.; Osorio, R.; Pastene, F.; Saavedra, F.; Pino, E. Capacitive sensors array for plantar pressure measurement insole fabricated with flexible PCB. In Proceedings of the 2018 40th Annual International Conference of the IEEE Engineering in Medicine and Biology Society (EMBC), Honolulu, HI, USA, 17–21 July 2018; pp. 4393–4396.
106. Rossetti, A.; Codeluppi, R.; Golfarelli, A.; Zagnoni, M.; Talamelli, A.; Tartagni, M. Design and characterization of polymeric pressure sensors for wireless wind sail monitoring. *Sens. Actuators A Phys.* **2011**, *167*, 162–170. [CrossRef]
107. Palasagaram, J.N.; Ramadoss, R. MEMS-capacitive pressure sensor fabricated using printed-circuit-processing techniques. *IEEE Sens. J.* **2006**, *6*, 1374–1375. [CrossRef]
108. Fuh, Y.K.; Ho, H.C. Highly flexible self-powered sensors based on printed circuit board technology for human motion detection and gesture recognition. *Nanotechnology* **2016**, *27*, 095401. [CrossRef]
109. Quoc, T.V.; Quoc, T.P.; Duc, T.C.; Bui, T.; Kikuchi, K.; Aoyagi, M. Capacitive sensor based on PCB technology for air bubble inside fluidic flow detection. In Proceedings of the SENSORS, 2014 IEEE, Valencia, Spain, 2–5 November 2014; pp. 237–240.
110. Quoc, T.V.; Dac, H.N.; Quoc, T.P.; Dinh, D.N.; Duc, T.C. A printed circuit board capacitive sensor for air bubble inside fluidic flow detection. *Microsyst. Technol.* **2015**, *21*, 911–918. [CrossRef]
111. Yan, D.; Yang, Y.; Hong, Y.; Liang, T.; Yao, Z.; Chen, X.; Xiong, J. Low-cost wireless temperature measurement: Design, manufacture, and testing of a PCB-based wireless passive temperature sensor. *Sensors* **2018**, *18*, 532. [CrossRef]
112. Aguilar, J.; Beadle, M.; Thompson, P.; Shelley, M. The microwave and RF characteristics of FR4 substrates. In Proceedings of the IEE Colloquium on Low Cost Antenna Technology (Ref. No. 1998/206) , 1998 IEEE, London, UK, 24 February 1998; pp. 2/1–2/6.
113. Ahn, C.H.; Kim, H.H.; Cha, J.M.; Kwon, B.H.; Ha, M.Y.; Park, S.H.; Jeong, J.H.; Kim, K.S.; Cho, J.R.; Son, C.M.; et al. Fabrication and Performance Evaluation of Temperature Sensor Matrix Using a Flexible Printed Circuit Board for the Visualization of Temperature Field. *J. Korean Soc. Vis.* **2010**, *7*, 17–21.
114. Fries, D.; Broadbent, H.; Steimle, G.; Ivanov, S.; Cardenas-Valencia, A.; Fu, J.; Weller, T.; Natarajan, S.; Guerra, L. PCB MEMS for environmental sensing systems. In Proceedings of the 31st Annual Conference of IEEE Industrial Electronics Society, IECON 2005, Raleigh, NC, USA, 6–10 November 2005; pp. 2352–2356.
115. Guijt, R.M.; Armstrong, J.P.; Candish, E.; Lefleur, V.; Percey, W.J.; Shabala, S.; Hauser, P.C.; Breadmore, M.C. Microfluidic chips for capillary electrophoresis with integrated electrodes for capacitively coupled conductivity detection based on printed circuit board technology. *Sens. Actuators B Chem.* **2011**, *159*, 307–313. [CrossRef]
116. Werner, F.T.; Dean, R.N. Characterising a PCB electrical conductivity sensor using electromagnetic simulation and a genetic algorithm. *IET Sci. Meas. Technol.* **2017**, *11*, 761–765. [CrossRef]
117. Liu, G.; Ho, C.; Slappey, N.; Zhou, Z.; Snelgrove, S.; Brown, M.; Grabinski, A.; Guo, X.; Chen, Y.; Miller, K.; et al. A wearable conductivity sensor for wireless real-time sweat monitoring. *Sens. Actuators B Chem.* **2016**, *227*, 35–42. [CrossRef]
118. Dean, R.N.; Guertal, E.A.; Newby, A.F. A low-cost environmental nitrate sensor. In Proceedings of the 2020 IEEE Green Technologies Conference (GreenTech), Oklahoma City, OK, USA, 1–3 April 2020; pp. 165–170.
119. Morais, F.; Carvalhaes-Dias, P.; Duarte, L.; Costa E.; Ferreira A.; Dias J.S. Fringing field capacitive smart sensor based on PCB technology for measuring water content in paper pulp. *J. Sens.* **2020**, *2020*, 3905804. [CrossRef]

120. Blaž, N.; Mišković, G.; Marić, A.; Damnjanović, M.; Radosavljević, G.; Živanov, L. Modeling and characterization of LC displacement sensor in PCB technology. In Proceedings of the 2012 35th International Spring Seminar on Electronics Technology, Bad Aussee, Austria, 9–13 May 2012; pp. 394–398.
121. Nguyen, N.; Huang, X.; Toh, K. Thermal flow sensor for ultra-low velocities based on printed circuit board technology. *Meas. Sci. Technol.* **2001**, *12*, 2131. [CrossRef]
122. Hasch, J.; Wostradowski, U.; Hellinger, R.; Mittelstrass, D. 77 GHz automotive radar sensor in low-cost PCB technology. In Proceedings of the 2011 8th European Radar Confererence, Manchester, UK, 12–14 October 2011; pp. 101–104.
123. Qiao, D.; Pang, G.K.; Mui, M.K.; Lam, D.C. A single-axis low-cost accelerometer fabricated using printed-circuit-board techniques. *IEEE Electron Device Lett.* **2009**, *30*, 1293–1295. [CrossRef]
124. Luque, A.; Flores, G.; Perdigones, F.; Medina, D.; Garcia, J.; Quero, J. Single axis accelerometer fabricated using printed circuit board techniques and laser ablation. *Sens. Actuators A Phys.* **2013**, *192*, 119–123. [CrossRef]
125. Rogers, J.; Ramadoss, R.; Ozmun, P.; Dean, R. A microelectromechanical accelerometer fabricated using printed circuit processing techniques. *J. Micromechan. Microeng.* **2007**, *18*, 015013. [CrossRef]
126. Lei, H.; Wen, Q.; Yu, F.; Zhou, Y.; Wen, Z. FR4-based electromagnetic scanning micromirror integrated with angle sensor. *Micromachines* **2018**, *9*, 214. [CrossRef] [PubMed]
127. Abgrall, P.; Gue, A. Lab-on-chip technologies: Making a microfluidic network and coupling it into a complete microsystem—A review. *J. Micromechan. Microeng.* **2007**, *17*, R15. [CrossRef]
128. Gupta, S.; Ramesh, K.; Ahmed, S.; Kakkar, V. Lab-on-chip technology: A review on design trends and future scope in biomedical applications. *Int. J. Bio-Sci. Bio-Technol.* **2016**, *8*, 311–322. [CrossRef]
129. Chow, A.W. Lab-on-Chip: Opportunities for chemical engineering. *Am. Inst. Chem. Eng. J.* **2002**, *48*, 1590. [CrossRef]
130. Moschou, D.; Tserepi, A. The lab-on-PCB approach: Tackling the μTAS commercial upscaling bottleneck. *Lab Chip* **2017**, *17*, 1388–1405. [CrossRef] [PubMed]
131. Perdigones, F. Lab-on-PCB and Flow Driving: A Critical Review. *Micromachines* **2021**, *12*, 175. [CrossRef]
132. Zhao, W.; Tian, S.; Huang, L.; Liu, K.; Dong, L. The review of Lab-on-PCB for biomedical application. *Electrophoresis* **2020**, *41*, 1433–1445. [CrossRef]
133. Ali, N.R.; Ahaitouf, A.; Abdullah, M.Z. Irreversible bonding techniques for the fabrication of a leakage-free printed circuit board-based lab-on-chip in microfluidic platforms—A review. *Meas. Sci. Technol.* **2021**, *32*, 052001.
134. Flores, G.; Aracil, C.; Perdigones, F.; Quero, J. Low consumption single-use microvalve for microfluidic PCB-based platforms. *J. Micromechan. Microeng.* **2014**, *24*, 065013. [CrossRef]
135. Perdigones, F.; Franco, E.; Salvador, B.; Flores, G.; Quero, J.M. Highly integrable microfluidic impulsion system for precise displacement of liquids on lab on PCBs. *J. Microelectromechan. Syst.* **2018**, *27*, 479–486. [CrossRef]
136. Flores, G.; Perdigones, F.; Aracil, C.; Quero, J. Pressurization method for controllable impulsion of liquids in microfluidic platforms. *Microelectron. Eng.* **2015**, *140*, 11–17. [CrossRef]
137. Perdigones, F.; Quero, J.M. Highly integrable and normally open microvalve for industrial thermoplastic-based lab on PCB. *Sens. Actuators A Phys.* **2019**, *300*, 111639. [CrossRef]
138. Perdigones, F.; Aracil, C.; Moreno, J.M.; Luque, A.; Quero, J.M. Highly integrable pressurized microvalve for portable SU-8 microfluidic platforms. *J. Microelectromechan. Syst.* **2013**, *23*, 398–405. [CrossRef]
139. Aracil, C.; Perdigones, F.; Moreno, J.M.; Luque, A.; Quero, J.M. Portable Lab-on-PCB platform for autonomous micromixing. *Microelectron. Eng.* **2015**, *131*, 13–18. [CrossRef]
140. Flores, G.; Aracil, C.; Perdigones, F.; Quero, J.M. Lab-protocol-on-PCB: Prototype of a laboratory protocol on printed circuit board using MEMS technologies. *Microelectron. Eng.* **2018**, *200*, 26–31. [CrossRef]
141. Aracil, C.; Quero, J.M.; Luque, A.; Moreno, J.M.; Perdigones, F. Pneumatic impulsion device for microfluidic systems. *Sens. Actuators A Phys.* **2010**, *163*, 247–254. [CrossRef]
142. Babikian, S.; Jinsenji, M.; Bachman, M.; Li, G. Surface Mount Electroosmotic Pump for Integrated Microfluidic Printed Circuit Boards. In Proceedings of the 2018 IEEE 68th Electronic Components and Technology Conference (ECTC), San Diego, CA, USA, 29 May–1 June 2018; pp. 498–503.
143. Gassmann, S.; Pagel, L.; Luque, A.; Perdigones, F.; Aracil, C. Fabrication of electroosmotic micropump using PCB and SU-8. In Proceedings of the 38th Annual Conference on IEEE Industrial Electronics Society, IECON 2012, Montreal, QC, Canada, 25–28 October 2012; pp. 3958–3961.
144. Luque, A.; Soto, J.M.; Perdigones, F.; Aracil, C.; Quero, J.M. Electroosmotic impulsion device for integration in PCB-MEMS. In Proceedings of the 2013 IEEE 9th Spanish Conference on Electron Devices, (CDE 2013), Valladolid, Spain, 12–14 February 2013; pp. 119–122.
145. Yi, Z.; Feng, H.; Zhou, X.; Shui, L. Design of an open electrowetting on dielectric device based on printed circuit board by using a parafilm m. *Front. Phys.* **2020**, *8*, 193. [CrossRef]
146. Boles, D.J.; Benton, J.L.; Siew, G.J.; Levy, M.H.; Thwar, P.K.; Sandahl, M.A.; Rouse, J.L.; Perkins, L.C.; Sudarsan, A.P.; Jalili, R.; et al. Droplet-based pyrosequencing using digital microfluidics. *Anal. Chem.* **2011**, *83*, 8439–8447. [CrossRef]
147. Pollack, M.G.; Pamula, V.K.; Srinivasan, V.; Eckhardt, A.E. Applications of electrowetting-based digital microfluidics in clinical diagnostics. *Expert Rev. Mol. Diagn.* **2011**, *11*, 393–407. [CrossRef]

148. Liu, R.H.; Yang, J.; Lenigk, R.; Bonanno, J.; Grodzinski, P. Self-contained, fully integrated biochip for sample preparation, polymerase chain reaction amplification, and DNA microarray detection. *Anal. Chem.* **2004**, *76*, 1824–1831. [CrossRef]
149. Schumacher, S.; Nestler, J.; Otto, T.; Wegener, M.; Ehrentreich-Förster, E.; Michel, D.; Wunderlich, K.; Palzer, S.; Sohn, K.; Weber, A.; et al. Highly-integrated lab-on-chip system for point-of-care multiparameter analysis. *Lab Chip* **2012**, *12*, 464–473. [CrossRef]
150. Kim, H.; Hwang, H.; Baek, S.; Kim, D. Design, fabrication and performance evaluation of a printed-circuit-board microfluidic electrolytic pump for lab-on-a-chip devices. *Sens. Actuators A Phys.* **2018**, *277*, 73–84. [CrossRef]
151. Mikhaylov, R.; Martin, M.S.; Dumcius, P.; Wang, H.; Wu, F.; Zhang, X.; Akhimien, V.; Sun, C.; Clayton, A.; Fu, Y.; et al. A reconfigurable and portable acoustofluidic system based on flexible printed circuit board for the manipulation of microspheres. *J. Micromechan. Microeng.* **2021**, *31*, 074003. [CrossRef]
152. Mikhaylov, R.; Wu, F.; Wang, H.; Clayton, A.; Sun, C.; Xie, Z.; Liang, D.; Dong, Y.; Yuan, F.; Moschou, D.; et al. Development and characterisation of acoustofluidic devices using detachable electrodes made from PCB. *Lab Chip* **2020**, *20*, 1807–1814. [CrossRef]
153. Lee, P.Y.; Costumbrado, J.; Hsu, C.Y.; Kim, Y.H. Agarose gel electrophoresis for the separation of DNA fragments. *J. Vis. Exp.* **2012**, *62*, e3923. [CrossRef]
154. Wuethrich, A.; Quirino, J.P. A decade of microchip electrophoresis for clinical diagnostics—A review of 2008–2017. *Anal. Chim. Acta* **2019**, *1045*, 42–66. [CrossRef]
155. Swerdlow, H.; Gesteland, R. Capillary gel electrophoresis for rapid, high resolution DNA sequencing. *Nucleic Acids Res.* **1990**, *18*, 1415–1419. [CrossRef]
156. Hempel, G. Strategies to improve the sensitivity in capillary electrophoresis for the analysis of drugs in biological fluids. *ELECTROPHORESIS Int. J.* **2000**, *21*, 691–698. [CrossRef]
157. Chang, Y.; You, H. A hybrid adhesive bonding of PMMA and PCB with an application on microchip electrophoresis. *Anal. Methods* **2019**, *11*, 1229–1236. [CrossRef]
158. Shadpour, H.; Hupert, M.L.; Patterson, D.; Liu, C.; Galloway, M.; Stryjewski, W.; Goettert, J.; Soper, S.A. Multichannel microchip electrophoresis device fabricated in polycarbonate with an integrated contact conductivity sensor array. *Anal. Chem.* **2007**, *79*, 870–878. [CrossRef] [PubMed]
159. Pethig, R. Dielectrophoresis: Status of the theory, technology, and applications. *Biomicrofluidics* **2010**, *4*, 022811. [CrossRef]
160. Park, K.; Suk, H.J.; Akin, D.; Bashir, R. Dielectrophoresis-based cell manipulation using electrodes on a reusable printed circuit board. *Lab Chip* **2009**, *9*, 2224–2229. [CrossRef]
161. Altomare, L.; Borgatti, M.; Medoro, G.; Manaresi, N.; Tartagni, M.; Guerrieri, R.; Gambari, R. Levitation and movement of human tumor cells using a printed circuit board device based on software-controlled dielectrophoresis. *Biotechnol. Bioeng.* **2003**, *82*, 474–479. [CrossRef]
162. Bhatt, G.; Kant, R.; Mishra, K.; Yadav, K.; Singh, D.; Gurunath, R.; Bhattacharya, S. Impact of surface roughness on dielectrophoretically assisted concentration of microorganisms over PCB based platforms. *Biomed. Microdevices* **2017**, *19*, 28. [CrossRef]
163. Vasilakis, N.; Moschou, D.; Carta, D.; Morgan, H.; Prodromakis, T. Long-lasting FR-4 surface hydrophilisation towards commercial PCB passive microfluidics. *Appl. Surf. Sci.* **2016**, *368*, 69–75. [CrossRef]
164. Rethmel, C.; Little, J.; Takashima, K.; Sinha, A.; Adamovich, I.; Samimy, M. Flow separation control using nanosecond pulse driven DBD plasma actuators. *Int. J. Flow Control* **2011**, *3*, 213–232. [CrossRef]
165. Neretti, G.; Ricchiuto, A.; Borghi, C. Measurement of the charge distribution deposited by an annular plasma synthetic jet actuator over a target surface. *J. Phys. D Appl. Phys.* **2018**, *51*, 324004. [CrossRef]
166. Rigit, A.R.H.; Lai, K.C.; Bong, D.B.L. Degradation of a dielectric barrier discharge plasma actuator. In Proceedings of the 2009 IEEE 9th International Conference on the Properties and Applications of Dielectric Materials, (ICPADM 2009), Harbin, China, 19–23 June 2009; pp. 569–572.
167. Lee, S.M.; Oh, I.Y.; Yook, J.G.; Hong, Y. Scattering characteristics of atmospheric pressure dielectric barrier discharge plasma. In Proceedings of the 2013 European Radar Conference, Nuremberg, Germany, 9–11 October 2013; pp. 555–558.
168. Rigit, A.R.H.; Ali, I.; Boon, T.C.; Chong, C.H. Effect of Number of Electrodes on Electrical Performance of Surface Dielectric Barrier Discharge Plasma Actuator. In Proceedings of the 2020 13th International UNIMAS Engineering Conference (EnCon), Kota Samarahan, Sarawak, Malaysia, 27–28 October 2020; pp. 1–4.
169. Neretti, G.; Seri, P.; Taglioli, M.; Shaw, A.; Iza, F.; Borghi, C.A. Geometry optimization of linear and annular plasma synthetic jet actuators. *J. Phys. D Appl. Phys.* **2016**, *50*, 015210. [CrossRef]
170. Shimizu, K.; Kristof, J.; Blajan, M.G. Applications of Dielectric Barrier Discharge Microplasma. In *Atmospheric Pressure Plasma from Diagnostics to Applications*; IntechOpen: London, UK, 2018.
171. Lammerink, T.; Spiering, V.; Elwenspoek, M.; Fluitman, J.; Van den Berg, A. Modular concept for fluid handling systems. A demonstrator micro analysis system. In Proceedings of the Ninth International Workshop on Micro Electromechanical Systems, San Diego, CA, USA, 11–15 February 1996; pp. 389–394.
172. Shamkhalichenar, H.; Bueche, C.J.; Choi, J.W. Printed Circuit Board (PCB) Technology for Electrochemical Sensors and Sensing Platforms. *Biosensors* **2020**, *10*, 159. [CrossRef]
173. Rose, D.P.; Ratterman, M.E.; Griffin, D.K.; Hou, L.; Kelley-Loughnane, N.; Naik, R.R.; Hagen, J.A.; Papautsky, I.; Heikenfeld, J.C. Adhesive RFID sensor patch for monitoring of sweat electrolytes. *IEEE Trans. Biomed. Eng.* **2014**, *62*, 1457–1465. [CrossRef]

174. Moschou, D.; Trantidou, T.; Regoutz, A.; Carta, D.; Morgan, H.; Prodromakis, T. Surface and electrical characterization of Ag/AgCl pseudo-reference electrodes manufactured with commercially available PCB technologies. *Sensors* **2015**, *15*, 18102–18113. [CrossRef] [PubMed]
175. Moreira, F.T.; Ferreira, M.J.M.; Puga, J.R.; Sales, M.G.F. Screen-printed electrode produced by printed-circuit board technology. Application to cancer biomarker detection by means of plastic antibody as sensing material. *Sens. Actuators B Chem.* **2016**, *223*, 927–935. [CrossRef]
176. Abedeen, Z.; Agarwal, P. Microwave sensing technique based label-free and real-time planar glucose analyzer fabricated on FR4. *Sens. Actuators A Phys.* **2018**, *279*, 132–139. [CrossRef]
177. Bag, S.; Pal, K. A PCB Based chemiresistive carbon dioxide sensor operating at room temperature under different relative humidity. *IEEE Trans. Nanotechnol.* **2019**, *18*, 1119–1128. [CrossRef]
178. Lall, P.; Narangaparambil, J.; Abrol, A.; Leever, B.; Marsh, J. Development of test protocols for the flexible substrates in wearable applications. In Proceedings of the 2018 17th IEEE Intersociety Conference on Thermal and Thermomechanical Phenomena in Electronic Systems (ITherm), San Diego, CA, USA, 29 May–1 June 2018; pp. 1120–1127.
179. Kim, C.S.; Yang, H.M.; Lee, J.; Lee, G.S.; Choi, H.; Kim, Y.J.; Lim, S.H.; Cho, S.H.; Cho, B.J. Self-powered wearable electrocardiography using a wearable thermoelectric power generator. *ACS Energy Lett.* **2018**, *3*, 501–507. [CrossRef]
180. Tao, X.; Huang, T.H.; Shen, C.L.; Ko, Y.C.; Jou, G.T.; Koncar, V. Bluetooth Low Energy-Based Washable Wearable Activity Motion and Electrocardiogram Textronic Monitoring and Communicating System. *Adv. Mater. Technol.* **2018**, *3*, 1700309. [CrossRef]
181. Windmiller, J.R.; Wang, J. Wearable electrochemical sensors and biosensors: A review. *Electroanalysis* **2013**, *25*, 29–46. [CrossRef]
182. Damiati, S.; Søpstad, S.; Peacock, M.; Akhtar, A.S.; Pinto, I.; Soares, R.; Russom, A. Flex Printed Circuit Board Implemented Graphene-Based DNA Sensor for Detection of SARS-CoV-2. *IEEE Sens. J.* **2021**, *21*, 13060–13067. [CrossRef]
183. Du, P.; O'Grady, G.; Egbuji, J.U.; Lammers, W.; Budgett, D.; Nielsen, P.; Windsor, J.; Pullan, A.; Cheng, L. High-resolution mapping of in vivo gastrointestinal slow wave activity using flexible printed circuit board electrodes: Methodology and validation. *Ann. Biomed. Eng.* **2009**, *37*, 839–846. [CrossRef] [PubMed]
184. Dutta, G.; Regoutz, A.; Moschou, D. Commercially fabricated printed circuit board sensing electrodes for biomarker electrochemical detection: The importance of electrode surface characteristics in sensor performance. *Proceedings* **2018**, *2*, 741.
185. Kim, K.Y.; Chang, H.; Lee, W.D.; Cai, Y.F.; Chen, Y.J. The influence of blood glucose meter resistance variation on the performance of a biosensor with a gold-coated circuit board. *J. Sens.* **2019**, *2019*, 5948182. [CrossRef]
186. Pittet, P.; Lu, G.N.; Galvan, J.M.; Ferrigno, R.; Blum, L.J.; Leca-Bouvier, B.D. PCB technology-based electrochemiluminescence microfluidic device for low-cost portable analytical systems. *IEEE Sens. J.* **2008**, *8*, 565–571. [CrossRef]
187. Pittet, P.; Lu, G.N.; Galvan, J.M.; Ferrigno, R.; Blum, L.J.; Leca-Bouvier, B. PCB-based integration of electrochemiluminescence detection for microfluidic systems. *Analyst* **2007**, *132*, 409–411. [CrossRef]
188. Kumar, M.; Nandeshwar, R.; Lad, S.B.; Megha, K.; Mangat, M.; Butterworth, A.; Knapp, C.W.; Knapp, M.; Hoskisson, P.A.; Corrigan, D.K.; et al. Electrochemical sensing of SARS-CoV-2 amplicons with PCB electrodes. *Sens. Actuators B Chem.* **2021**, *343*, 130169. [CrossRef]
189. Nandeshwar, R.; Kumar, M.S.; Kondabagil, K.; Tallur, S. Electrochemical Immunosensor Platform Using Low-Cost ENIG PCB Finish Electrodes: Application For SARS-CoV-2 Spike Protein Sensing. *IEEE Access* **2021**, *9*, 154368–154377. [CrossRef]
190. Bonilla, D.; Mallen, M.; De La Rica, R.; Fernandez-Sanchez, C.; Baldi, A. Electrical readout of protein microarrays on regular glass slides. *Anal. Chem.* **2011**, *83*, 1726–1731. [CrossRef]
191. Yildirim, B.; Senveli, S.U.; Gajasinghe, R.W.; Tigli, O. Surface Acoustic Wave Viscosity Sensor with Integrated Microfluidics on a PCB Platform. *IEEE Sens. J.* **2018**, *18*, 2305–2312. [CrossRef]
192. Coyle, S.; Lau, K.T.; Moyna, N.; O'Gorman, D.; Diamond, D.; Di Francesco, F.; Costanzo, D.; Salvo, P.; Trivella, M.G.; De Rossi, D.E.; et al. BIOTEX—Biosensing textiles for personalised healthcare management. *IEEE Trans. Inf. Technol. Biomed.* **2010**, *14*, 364–370. [CrossRef]
193. Zhang, X.Y.; Li, Z.Y.; Zhang, Y.; Zang, X.Q.; Ueno, K.; Misawa, H.; Sun, K. Bacterial concentration detection using a PCB-based contactless conductivity sensor. *Micromachines* **2019**, *10*, 55. [CrossRef] [PubMed]
194. Jaanus, M.; Udal, A.; Kukk, V.; Umbleja, K.; Gorbatsova, J.; Molder, L. Improved C5D electronic realization of conductivity detector for capillary electrophoresis. *Elektron. Elektrotech.* **2016**, *22*, 29–32. [CrossRef]
195. Cortina, M.; Esplandiu, M.; Alegret, S.; Del Valle, M. Urea impedimetric biosensor based on polymer degradation onto interdigitated electrodes. *Sens. Actuators B Chem.* **2006**, *118*, 84–89. [CrossRef]
196. Cui, H.; Xiong, X.; Gao, B.; Chen, Z.; Luo, Y.; He, F.; Deng, S.; Chen, L. A novel impedimetric biosensor for detection of lead (II) with low-cost interdigitated electrodes made on PCB. *Electroanalysis* **2016**, *28*, 2000–2006. [CrossRef]
197. Dudala, S.; Srikanth, S.; Dubey, S.K.; Javed, A.; Goel, S. Rapid Inkjet-Printed Miniaturized Interdigitated Electrodes for Electrochemical Sensing of Nitrite and Taste Stimuli. *Micromachines* **2021**, *12*, 1037. [CrossRef]
198. Dean, R.N.; Rane, A.K.; Baginski, M.E.; Richard, J.; Hartzog, Z.; Elton, D.J. A capacitive fringing field sensor design for moisture measurement based on printed circuit board technology. *IEEE Trans. Instrum. Meas.* **2011**, *61*, 1105–1112. [CrossRef]
199. Novickij, V.; Tabasnikov, A.; Smith, S.; Grainys, A.; Novickij, J. Analysis of planar circular interdigitated electrodes for electroporation. *IETE Tech. Rev.* **2015**, *32*, 196–203. [CrossRef]

200. Kassanos, P.; Seichepine, F.; Kassanos, I.; Yang, G.Z. Development and Characterization of a PCB-Based Microfluidic YChannel. In Proceedings of the 2020 42nd Annual International Conference of the IEEE Engineering in Medicine & Biology Society (EMBC), Montreal, QC, Canada, 20–24 July 2020; pp. 5037–5040.
201. Papadopoulos, V.E.; Kokkoris, G.; Kefala, I.N.; Tserepi, A. Comparison of continuous-flow and static-chamber μPCR devices through a computational study: The potential of flexible polymeric substrates. *Microfluid. Nanofluidics* **2015**, *19*, 867–882. [CrossRef]
202. Kaprou, G.D.; Papadopoulos, V.; Papageorgiou, D.P.; Kefala, I.; Papadakis, G.; Gizeli, E.; Chatzandroulis, S.; Kokkoris, G.; Tserepi, A. Ultrafast, low-power, PCB manufacturable, continuous-flow microdevice for DNA amplification. *Anal. Bioanal. Chem.* **2019**, *411*, 5297–5307. [CrossRef]
203. Kaprou, G.; Papadakis, G.; Papageorgiou, D.; Kokkoris, G.; Papadopoulos, V.; Kefala, I.; Gizeli, E.; Tserepi, A. Miniaturized devices for isothermal DNA amplification addressing DNA diagnostics. *Microsyst. Technol.* **2016**, *22*, 1529–1534. [CrossRef]
204. Talebi, F.; Ghafoorifard, H.; Ghafouri-Fard, S.; Jahanshahi, A. A straight-forward, inexpensive, low power continuous-flow μPCR chip using PCB-based heater electrodes with uniform temperature distribution. *Sens. Actuators A Phys.* **2022**, *333*, 113220. [CrossRef]
205. Gassmann, S.; Götze, H.; Hinze, M.; Mix, M.; Flechsig, G.U.; Pagel, L. PCB based DNA detection chip. In Proceedings of the 38th Annual Conference on IEEE Industrial Electronics Society, IECON 2012, Montreal, QC, Canada, 25–28 October 2012; pp. 3982–3986.
206. Jolly, P.; Rainbow, J.; Regoutz, A.; Estrela, P.; Moschou, D. A PNA-based Lab-on-PCB diagnostic platform for rapid and high sensitivity DNA quantification. *Biosens. Bioelectron.* **2019**, *123*, 244–250. [CrossRef] [PubMed]
207. Gassmann, S.; Trozjuk, A.; Singhal, J.; Schuette, H.; Miranda, M.L.; Zielinski, O. PCB based micro fluidic system for thermal cycling of seawater samples. In Proceedings of the 2015 IEEE International Conference on Industrial Technology (ICIT 2015), Seville, Spain, 17–19 March 2015; pp. 3365–3369.
208. Fu, Y.; Yuan, Q.; Guo, J. Lab-on-PCB-based micro-cytometer for circulating tumor cells detection and enumeration. *Microfluid. Nanofluidics* **2017**, *21*, 20. [CrossRef]
209. Guo, J.; Li, H.; Chen, Y.; Kang, Y. A microfluidic impedance cytometer on printed circuit board for low cost diagnosis. *IEEE Sens. J.* **2013**, *14*, 2112–2117. [CrossRef]
210. Shi, D.; Guo, J.; Chen, L.; Xia, C.; Yu, Z.; Ai, Y.; Li, C.M.; Kang, Y.; Wang, Z. Differential microfluidic sensor on printed circuit board for biological cells analysis. *Electrophoresis* **2015**, *36*, 1854–1858. [CrossRef]
211. Zhao, Y.; Zhang, W. Biophysical measurement of red blood cells by laboratory on print circuit board chip. *Electrophoresis* **2019**, *40*, 1140–1143. [CrossRef]
212. Multichannel Systems. Available online: https://www.multichannelsystems.com/ (accessed on 13 February 2022).
213. Urbano-Gámez, J.D.; Valdés-Sánchez, L.; Aracil, C.; de la Cerda, B.; Perdigones, F.; Plaza Reyes, Á.; Díaz-Corrales, F.J.; Relimpio López, I.; Quero, J.M. Biocompatibility Study of a Commercial Printed Circuit Board for Biomedical Applications: Lab-on-PCB for Organotypic Retina Cultures. *Micromachines* **2021**, *12*, 1469. [CrossRef]
214. Ayanda Biosystems SA. Available online: https://alascience.com/products/pdf/MEA_Product_Catalog.pdf (accessed on 13 February 2022).
215. Burdallo, I.; Jimenez-Jorquera, C.; Fernández-Sánchez, C.; Baldi, A. Integration of microelectronic chips in microfluidic systems on printed circuit board. *J. Micromechan. Microeng.* **2012**, *22*, 105022. [CrossRef]
216. Carroll, A.E.; Marrero, D.G.; Downs, S.M. The HealthPia GlucoPack™ diabetes phone: A usability study. *Diabetes Technol. Ther.* **2007**, *9*, 158–164. [CrossRef]
217. Cafazzo, J.A.; Casselman, M.; Hamming, N.; Katzman, D.K.; Palmert, M.R. Design of an mHealth app for the self-management of adolescent type 1 diabetes: A pilot study. *J. Med. Internet Res.* **2012**, *14*, e2058. [CrossRef]
218. Azzarelli, J.M.; Mirica, K.A.; Ravnsbæk, J.B.; Swager, T.M. Wireless gas detection with a smartphone via rf communication. *Proc. Natl. Acad. Sci. USA* **2014**, *111*, 18162–18166. [CrossRef] [PubMed]
219. Krorakai, K.; Klangphukhiew, S.; Kulchat, S.; Patramanon, R. Smartphone-based NFC potentiostat for wireless electrochemical sensing. *Appl. Sci.* **2021**, *11*, 392. [CrossRef]
220. Xu, G.; Zhang, Q.; Lu, Y.; Liu, L.; Ji, D.; Li, S.; Liu, Q. Passive and wireless near field communication tag sensors for biochemical sensing with smartphone. *Sens. Actuators B Chem.* **2017**, *246*, 748–755. [CrossRef]
221. Teengam, P.; Siangproh, W.; Tontisirin, S.; Jiraseree-amornkun, A.; Chuaypen, N.; Tangkijvanich, P.; Henry, C.S.; Ngamrojanavanich, N.; Chailapakul, O. NFC-enabling smartphone-based portable amperometric immunosensor for hepatitis B virus detection. *Sens. Actuators B Chem.* **2021**, *326*, 128825. [CrossRef]
222. Shi, J.; Qing, X.; Chen, Z.N.; Goh, C.K. Electrically large dual-loop antenna for UHF near-field RFID reader. *IEEE Trans. Antennas Propag.* **2012**, *61*, 1019–1025. [CrossRef]
223. Maddela, M.; Ramadoss, R.; Lempkowski, R. A MEMS-based tunable coplanar patch antenna fabricated using PCB processing techniques. *J. Micromechan. Microeng.* **2007**, *17*, 812. [CrossRef]
224. Nawaz, M.I.; Huiling, Z.; Nawaz, M.S.S.; Zakim, K.; Zamin, S.; Khan, A. A review on wideband microstrip patch antenna design techniques. In Proceedings of the 2013 International Conference on Aerospace Science & Engineering (ICASE), Islamabad, Pakistan, 21–23 August 2013; pp. 1–8.

225. Lamminen, A.; Säily, J.; Ala-Laurinaho, J.; de Cos, J.; Ermolov, V. Patch antenna and antenna array on multilayer high-frequency PCB for D-band. *IEEE Open J. Antennas Propag.* **2020**, *1*, 396–403. [CrossRef]
226. Dastranj, A.; Abiri, H. Bandwidth enhancement of printed E-shaped slot antennas fed by CPW and microstrip line. *IEEE Trans. Antennas Propag.* **2010**, *58*, 1402–1407. [CrossRef]
227. Tripathi, N.M. Miniaturized meander PCB antenna for 433MHz. In Proceedings of the 2015 IEEE Applied Electromagnetics Conference (AEMC) Guwahati, Assam, India, 18–21 December, 2015; pp. 1–2.
228. Cheung, C.Y.; Yuen, J.S.; Mung, S.W. Miniaturized printed inverted-F antenna for internet of things: A design on PCB with a meandering line and shorting strip. *Int. J. Antennas Propag.* **2018**, *2018*, 5948182.
229. Kumar, A.; Saravanakumar, M.; Raghavan, S. Dual-frequency SIW-based cavity-backed antenna. *AEU Int. J. Electron. Commun.* **2018**, *97*, 195–201. [CrossRef]
230. Rais, N.; Soh, P.J.; Malek, F.; Ahmad, S.; Hashim, N.; Hall, P. A review of wearable antenna. In Proceedings of the 2009 Loughborough Antennas & Propagation Conference, Loughborough, UK, 16–17 November 2009; pp. 225–228.
231. Lee, Y.; Paik, J.; Lee, S. Development of Miniature Quad SAW filter bank based on PCB substrate. In Proceedings of the 2007 IEEE International Frequency Control Symposium Joint with the 21st European Frequency and Time Forum, Geneva, Switzerland, 29 May–1 June 2007; pp. 146–149.
232. Li, C.W.; Cheung, C.N.; Yang, J.; Tzang, C.H.; Yang, M. PDMS-based microfluidic device with multi-height structures fabricated by single-step photolithography using printed circuit board as masters. *Analyst* **2003**, *128*, 1137–1142. [CrossRef]
233. Tu, J.; Qiao, Y.; Feng, H.; Li, J.; Fu, J.; Liang, F.; Lu, Z. PDMS-based microfluidic devices using commoditized PCBs as masters with no specialized equipment required. *RSC Adv.* **2017**, *7*, 31603–31609. [CrossRef]
234. Chee, P.S.; Arsat, R.; Hashim, U.; Abdul Rahim, R.; Leow, P.L. Micropump pattern replication using Printed Circuit Board (PCB) technology. *Mater. Manuf. Process.* **2013**, *28*, 702–706.
235. Villegas, J.; Perdigones, F.; Moreno, J.; Quero, J. Towards a low-cost production of monodispersed microbubbles using PCB-MEMS technology. In Proceedings of the 8th Spanish Conference on Electron Devices, CDE'2011, Palma de Mallorca, Spain, 8–11 February 2011; pp. 1–4.
236. Li, J.; Wang, Y.; Dong, E.; Chen, H. USB-driven microfluidic chips on printed circuit boards. *Lab Chip* **2014**, *14*, 860–864. [CrossRef]
237. Perdigones, F.; Moreno, J.; Luque, A.; Quero, J. Characterisation of the fabrication process of freestanding SU-8 microstructures integrated in printing circuit board in microelectromechanical systems. *Micro Nano Lett.* **2010**, *5*, 7–13. [CrossRef]
238. Perdigones, F.; Moreno, J.M.; Luque, A.; Aracil, C.; Quero, J.M. Safety valve in PCB-MEMS technology for limiting pressure in microfluidic applications. In Proceedings of the 2010 IEEE International Conference on Industrial Technology, (ICIT 2010), Viña del Mar, Chile, 14–17 March 2010; pp. 1558–1561.
239. Aracil, C.; Perdigones, F.; Riveros, J.A.; Bogado, B.; Quero, J.M. Pneumatic free piston fabricated in SU-8 for MEMS applications. In Proceedings of the 8th Spanish Conference on Electron Devices, CDE'2011, Palma de Mallorca, Spain, 8–11 February 2011; pp. 1–4.
240. Gassmann, S.; Luque, A.; Perdigones, F.; Quero, J.M.; Pagel, L. Sensor structures generated with combination of SU8 and PCBMEMS. In Proceedings of the 39th Annual Conference of the IEEE Industrial Electronics Society, IECON 2013, Vienna, Austria, 10–13 November 2013; pp. 108–112.
241. Chin, L.Y.; Chong, K.K. Study of high power light emitting diode (LED) lighting system in accelerating the growth rate of Lactuca sativa for indoor cultivation. *Int. J. Phys. Sci.* **2012**, *7*, 1773–1781.
242. Bodart, D.; Dehning, B.; Levasseur, S.; Pacholek, P.; Rakai, A.; Sapinski, M.; Satou, K.; Schneider, G.; Steyart, D.; Storey, J. Development of an ionization profile monitor based on a pixel detector for the CERN Proton Synchrotron. In Proceedings of the IBIC2015, Melbourne, Australia, 13–17 September 2015; pp. 13–17.
243. Parrain, F.; Charlot, B.; Galy, N.; Courtois, B. CMOS-compatible micromachined tactile fingerprint sensor. In *Design, Test, Integration, and Packaging of MEMS/MOEMS 2002*; International Society for Optics and Photonics: Bellingham, DC, USA 2002; Volume 4755, pp. 568–575.
244. Mintenbeck, J.; Estaña, R.; Woern, H. Design of a modular, flexible instrument with integrated DC-motors for minimal invasive robotic surgery. In Proceedings of the 2013 IEEE/ASME International Conference on Advanced Intelligent Mechatronics, Wollongong, Australia, 9–12 June 2013; pp. 1249–1254.
245. Nikshoar, M.S.; Khosravi, S.; Jahangiri, M.; Zandi, A.; Miripour, Z.S.; Bonakdar, S.; Abdolahad, M. Distinguishment of populated metastatic cancer cells from primary ones based on their invasion to endothelial barrier by biosensor arrays fabricated on nanoroughened poly (methyl methacrylate). *Biosens. Bioelectron.* **2018**, *118*, 51–57. [CrossRef] [PubMed]
246. Salvador, B.; Franco, E.; Perdigones, F.; Quero, J.M. Fabrication process for inexpensive, biocompatible and transparent PCBs. Application to a flow meter. *Microelectron. Eng.* **2017**, *173*, 6–12. [CrossRef]
247. Kimtan, T.; Thupmongkol, J.; Williams, J.C.; Thongpang, S. Printable and transparent micro-electrocorticography (μECoG) for optogenetic applications. In Proceedings of the 2014 36th Annual International Conference of the IEEE Engineering in Medicine and Biology Society, Chicago, IL, USA, 26–30 August 2014; pp. 482–485.

Article

3D Printed PCB Microfluidics

Stefan Gassmann *, Sathurja Jegatheeswaran, Till Schleifer, Hesam Arbabi and Helmut Schütte

Department of Engineering, Jade University of Applied Sciences, 26389 Wilhelmshaven, Germany; sathurja.jegatheeswaran@jstudent.ade-hs.de (S.J.); till.schleifer@student.jade-hs.de (T.S.); hesam.arbabi@student.jade-hs.de (H.A.); helmut.schuette@jade-hs.de (H.S.)

* Correspondence: stefan.gassmann@jade-hs.de; Tel.: +49-4421-985-2353

Abstract: The combination of printed circuit boards (PCB) and microfluidics has many advantages. The combination of electrodes, sensors and electronics is needed for almost all microfluidic systems. Using PCBs as a substrate, this integration is intrinsic. Additive manufacturing has become a widely used technique in industry, research and by hobbyists. One very promising rapid prototype technique is vat polymerization with an LCD as mask, also known as masked stereolithography (mSLA). These printers are available with resolutions down to 35 μm, and they are affordable. In this paper, a technology is described which creates microfluidics on a PCB substrate using an mSLA printer. All steps of the production process can be carried out with commercially available printers and resins: this includes the structuring of the copper layer of the PCB and the buildup of the channel layer on top of the PCB. Copper trace dimensions down to 100 μm and channel dimensions of 800 μm are feasible. The described technology is a low-cost solution for combining PCBs and microfluidics.

Keywords: microfluidics; PCB; rapid prototyping; 3D printing; PCB-MEMS

Citation: Gassmann, S.; Jegatheeswaran, S.; Schleifer, T.; Arbabi, H.; Schütte, H. 3D Printed PCB Microfluidics. *Micromachines* **2022**, *13*, 470. https://doi.org/10.3390/mi13030470

Academic Editor: Aiqun Liu

Received: 25 February 2022
Accepted: 17 March 2022
Published: 19 March 2022

1. Introduction

Microfluidics has a lot of advantages. The usage of low volumes, the fast and well-controlled chemical reactions, the portability and the effects that appear at the micro-scale meet the needs of a huge range of applications. At the beginning of microfluidics research, a sophisticated micro-technologies lab was needed. The main technologies of the semi-conductor industry were used to generate the needed cavities. During the course of microfluidic technology development, many other technologies were developed, such as micro-machining, soft lithography, etc. [1–3]. One technological possibility combines printed circuit boards (PCBs) and micro-channels [4]. The advantage of this approach is the easy combination of electrodes, electronics and fluidic channels, as well as the well-established mass production of PCBs. The combination of electrodes and fluidics is of interest in all applications where an electric connection to the fluid is needed, e.g., electrochemical measurements, impedance spectroscopy or conductivity measurements. The combination of electronics and fluidics is also of interest in all applications where sensors and actuators that have electric connections need to be in contact with the fluid (e.g., thermal actuators by joule heating, electromagnetic, electrostatic, electrokinetic and electrowetting actuators, photometric sensors with LEDs and photo diodes and many more).

Additive manufacturing, also known as 3D printing, has also been used for the creation of microfluidics [5]. One very promising technology is masked stereolithography (mSLA), where photosensitive resins are polymerized by light activation. Many different resins with different properties (even biocompatibility) are on the market. A lot of affordable mSLA printers are available with typical x/y resolutions down to 35 μm. The most advanced microfluidic use of resin-based 3D printing is presented by Gong et al. Using customized hardware and resin, 18 μm × 20 μm-sized cavities are feasible [6].

In order to facilitate the fabrication of, and to disseminate the advantages of, microfluidics, the usage of commercially available mSLA printers and resins is of interest. This will enable the creation of microfluidic devices with low investments in lab infrastructure.

The combination of 3D printing and PCBs for fluidics was reported in [7]. Cabrera-López et al. used a 3D printed well and a commercially produced PCB for impedance spectroscopy measurements. The 3D printed part was made by an FDM (fused deposition modelling) printer, and contains no closed cavities. Both parts were detachable.

In this paper, a technology for the creation of microfluidics on PCB substrates using commercially available resins and mSLA printers is shown. The demonstrated technology includes the structuring of the copper layer of the PCB and the buildup of a channel layer, all using a commercially available mSLA printer. This technology can be used to create microfluidic designs in a simple laboratory environment in a short time. In time-limited education events such as summer schools the technique presented here will enable students to build their own designs.

In this paper, a microfluidic chip for the measurement of salinity using conductivity measurement is demonstrated. The chip contains a water-filled cavity and electrodes to perform the conductivity measurement. The production process and the first results of the chip test are demonstrated.

2. Materials and Methods

First the technology of the combination of PCBs and microfluidic channels is described. Second the example system that is created is shown in detail.

2.1. Technology

The 3D printer used here is a Sonic Mini 4k (Phrozen, phrozen3d.com). It is a commercially available mSLA printer with a resolution of 35 μm in x- and y-, and 10 μm in the z-direction. The build area is 13.4 cm × 7.5 cm × 13 cm. Other printer models can also be used. The only modification made was a holder for PCBs with the size of 50 mm × 50 mm.

Figure 1 shows the build plate of the used 3D printer with a mounted PCB. An aluminum frame with alignment points (highlighted in Figure 1) is added to the build plate. The alignment points are circular-shaped touchpoints created by milling in the opening for the PCB. With this manual alignment, an accuracy of 100 μm is feasible.

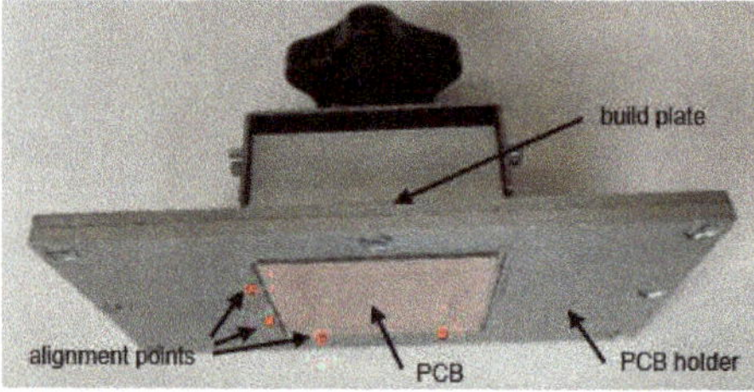

Figure 1. Printer build plate with holder for a PCB; the marked areas are alignment posts.

The PCB holder is the only modification needed to use the technology presented here.

All technology steps are pictured in Figure 2. The technology starts with the structuring of the copper (Figure 2a). For this, a single-sided PCB (1.5 mm FR4, 35 μm copper), coated with photosensitive resist, is used (presensitized boards, Bungard Elektronik & Co KG), and is cut into 50 mm × 50 mm pieces. The PCB is mounted with double-sided stick tape in the holder (Figure 1); after this, the Z-axis calibration takes place. As a printing file, the negative layout of the copper traces is loaded to the printer. The resist film needs to be exposed for 20 min with the used printer. After this, the exposed layer needs to be developed in the Bungard developer (article number: 72130–01, Bungard Elektronik & Co KG) for 5 min. The mask for etching the copper is now ready. For etching, a solution of sodium persulfate (ETCHANT 400 g, PROMA) can be used. The PCB used here was etched in an etching machine (AETZGERAET1, PROMA) at 40 °C for 10 min. The proposed procedure leads to the best results, with copper trace width down to 100 μm. The usage of the 3D printing resin as etch resist was also tested. It worked down to a copper trace width of 300 μm.

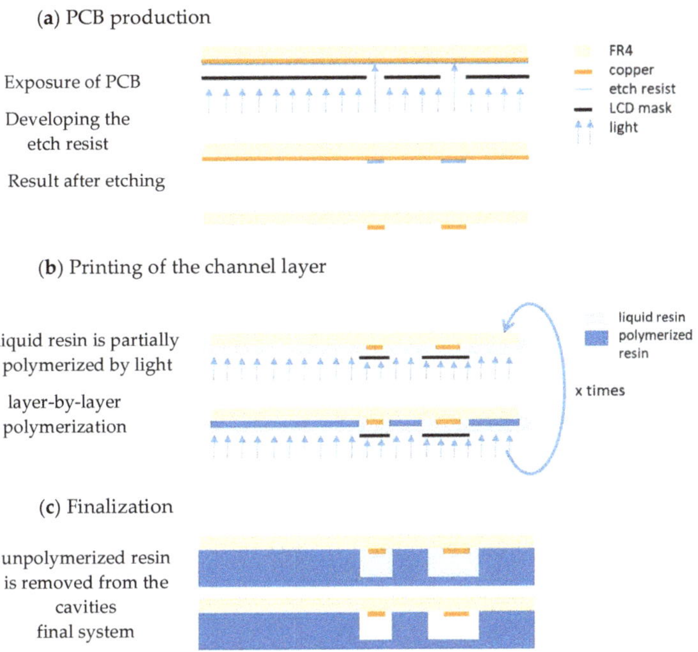

Figure 2. Technology steps: (**a**) PCB production—the etch resist is exposed by the 3D printer after the copper layer is etched as usual; (**b**) printing channel layer—the 3D printing process is carried out on the structured PCB; (**c**) finalization—unpolymerized resin must be removed from the cavities.

The second step is the printing of the channel layer on the structured PCB (Figure 2b). Right before printing the channel layer, the PCB must be cleaned with isopropanol. The so-prepared PCB is mounted again in the holder on the printer build plate using double-sided sticky tape. The same orientation as for the exposure of the etch resist should be used. This ensures the alignment of the copper trace with the channels. In preparation, before the print can start, Z-axis calibration must be performed with the mounted PCB. Now the channel layer can be printed directly on the PCB. For this purpose, the vat must be installed and filled with the resin. In addition, the printer file for the channel structure must be transferred on the printer and selected for printing. In this study, the color mix resin basic from 3Djake (niceshops GmbH, Paldau, Austria) was used. This resin needs longer exposure times than other resins; however, this longer exposure time leads to less light bleed and less over-exposure for the base layers, which makes it suitable for printing small cavities. The settings are: 4 bottom layers, 80 s exposure time for the bottom layer, 6.5 s exposure time for normal layers. The *z*-axis resolution used was 50 μm. As usual for the mSLA, 3D printing of the layer-by-layer buildup of the designed structure takes place at this step.

After printing the channel structures, the print must be cleaned (Figure 2c). All uncured resin must be washed off before final curing. For this purpose, the wash-and-clean station CW1S (Prusa Research a.s., Prague, Czech republic) was used. Cleaning was performed in isopropanol for 3 min. After the outside cleaning, the remaining resin inside the cavities must be removed. For this, a syringe filled with isopropanol was used, followed by cleaning with pressurized air. After cleaning, the final curing takes place, also in the CW1S. The curing program was run for 3 min.

The total procedure can be performed in less than three hours. This makes this technology suitable for time-limited education courses.

2.2. Example (Salinity Sensor)

The usage of PCBs as substrate for microfluidics has the advantage that the integration of electrodes and electronics in the microfluidic system is intrinsic. Using this approach, chip electrophoresis, conductivity measurements, spectrophotometry and many other applications can be built. Here, the conductivity measurement of seawater for the measurement of salinity is chosen as an example.

The system was designed using the 3D CAD program Fusion360® (Autodesk Inc.), which gives the possibility of co-designing the PCB and the channel. This has the advantage that the positioning of the electrodes and electronics and the channels can be designed properly in one tool. The steps for an electro-mechanical co-design are described in the Fusion360® documentation [8]. Figure 3 shows the CAD view and one realized prototype.

(**a**) CAD view of the system

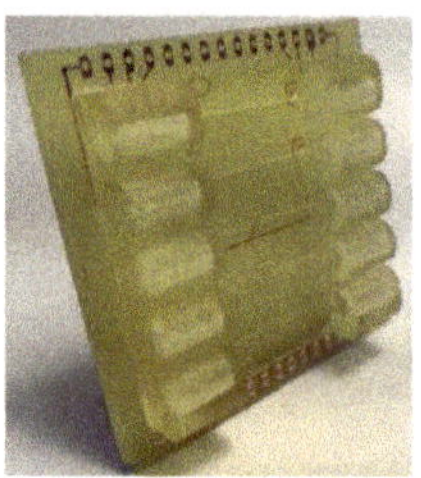

(**b**) photo of a prototype

Figure 3. The view in the CAD system and one realized prototype: (**a**) CAD view of the combined design—the PCB with the copper traces is in the back; the channels and the connectors are opaque gray; (**b**) photo of one realized prototype—the channel material is opaque for better visibility of the fluid.

The example used here consists of 5 different channels with different electrode arrangements for measuring the conductivity of the water in the channels. The different arrangements are for the demonstration of the 2-probe, the 4-probe and the guard electrode measurement. Closed cavities are useful in conductivity measurements to build a well-defined current path and prevent external fields. The system was built to demonstrate the salinity measurement to marine science students.

In the experiment, artificial seawater was created using sea salt and de-ionized water. The conductivity of the water was 55.3 mS/cm; it was measured with a GMH 3400 conductivity meter (Greisinger Messtechnik GmbH). The measurement of the resistance of the seawater in the channels was performed by an LCR meter HM8018 (Hameg instruments) at 25 kHz measurement frequency. The seawater was filled in the channels by 1 mL plastic syringes.

3. Results

3.1. Technology Results

Using the described technology systems containing electrodes, electronics and microfluidic channels are feasible. Only commercially available tools and consumables were used. The structure sizes were measured with a measuring microscope prior to the functional tests of the system. The minimal structure sizes of the 35 µm-thick copper traces are 100 µm. The cavity width (x/y resolution of the printer) in the presented system was 800 µm, which is appropriate for the demonstrated example but should be reduced in the future. Gong et al. concluded that a cavity dimension of 4 times the x/y resolution (pixel size) is feasible [6]. This is only possible using customized hardware and resin. The height of the cavity (z-direction of the printer) was measured by filling the cavity with a defined volume of water. Assuming a rectangular shape of the cavity, heights of about 200 μm were calculated. This was one-fifth of the designed height. The reason is that the entrapped resin in the cavity is exposed to a light dose with every layer above the cavity. Finally, this leads to an unwanted polymerization of the upper layer in the cavity. This effect should be taken into account when designing the cavity height.

An influence of the different optical properties of the copper and the FR4 was not observed during the production of the structure presented here. It may become apparent when the feature size of the channels becomes smaller and the adjacent copper areas become larger.

The problem of creating cavities with mSLA printers is that the enclosed resin that should stay not polymerized becomes exposed by stray light (this effect is also named light bleeding). This effect depends on the following factors: 1. Polymerization depth of the resin. A good description of the effects leading to this behavior can be found in [9]; 2. Design. Lower number of layers above the cavity and bigger inner dimensions of the cavity overcome the problem of polymerized encapsulated resin.

3.2. Conductivity Measurement Results

The results for the conductivity measurement are shown here only for the four-probe channel; the measurements of the other electrode arrangements agree with the results presented here.

With the calculated cross-section of the four-probe channel and the conductivity of the salt water, the resistance was calculated as 19.9 kOhm. The measured value with the HM 8018 was 20.9 kOhm (@25 kHz). The reason for the difference in the values could be the non-rectangular cross-section of the channel.

Using this microfluidic chip, the principle of salinity measurement by direct contact with the water can be demonstrated. The difference in electrode arrangements, the influence of the outside electric path and the usage of a guard electrode can be shown. With the system shown here, students will gain experience with the electric current field and with measuring salinity.

4. Conclusions

Microfluidics is no longer a domain of microtechnology laboratories with expensive equipment. If a cavity width of several hundreds of micrometers is enough, it is even possible to build microfluidic systems with a low-cost mSLA 3D printer and commercially available resins. The presented technology shows also that the combination of electrodes, electronics and microfluidic channels can be produced using the low-cost mSLA technique. The novelty of the technology is the usage of a PCB as a substrate. PCBs usually carry electronics and realize their wiring. When using PCBs as a substrate for microfluidics, the combination of electronics and fluidics is intrinsic. Combined electronic/microfluidic systems can integrate the electric connection to the fluid easily. The evaluation electronics can be located close to the sensors, and the system becomes compact and reliable due to the lack of wiring.

The technology shown here is a very good option for rapid prototyping and for educational purposes because of the low investments and the fast results. For mass production of PCB-based microfluidics, other technologies, such as the combination of hot embossed thermoplastic channels with mass-produced PCBs, are suitable.

With this technology, the structuring of the copper clad is possible in structure sizes down to 100 μm. The realized cavity width above the copper layer was 800 μm. Further research was will be carried out to minimize the feasible cavity size.

As an example, a microfluidic conductivity sensor chip was realized. Several electrode arrangements were demonstrated. The principle of salinity measurements using direct contact with the water, different electrode arrangements, such as the four-probe measurement, and the influence of the outside current path were demonstrated using the presented microfluidic chip.

Author Contributions: Conceptualization, S.G. and H.S.; investigation and experiments, S.J., T.S. and H.A.; writing—original draft preparation, S.G.; supervision, S.G.; project administration, S.G., All authors have read and agreed to the published version of the manuscript.

Funding: This research received no external funding.

Conflicts of Interest: The authors declare no conflict of interest.

References

1. Whitesides, G.M. The origins and the future of microfluidics. *Nature* **2006**, *442*, 368–373. [CrossRef] [PubMed]
2. Niculescu, A.-G.; Chircov, C.; Bîrcă, A.C.; Grumezescu, A.M. Fabrication and Applications of Microfluidic Devices: A Review. *Int. J. Mol. Sci.* **2021**, *22*, 2011. [CrossRef] [PubMed]
3. Abgrall, P.; Gue, A.-M. Lab-on-chip technologies: Making a microfluidic network and coupling it into a complete microsystem—A review. *J. Micromech. Microeng.* **2007**, *17*, R15–R49. [CrossRef]
4. Moschou, D.; Tserepi, A. The lab-on-PCB approach: Tackling the μTAS commercial upscaling bottleneck. *Lab Chip* **2017**, *17*, 1388–1405. [CrossRef] [PubMed]
5. Bhattacharjee, N.; Urrios, A.; Kang, S.; Folch, A. The upcoming 3D-printing revolution in microfluidics. *Lab Chip* **2016**, *16*, 1720–1742. [CrossRef] [PubMed]
6. Gong, H.; Bickham, B.P.; Woolley, A.T.; Nordin, G.P. Custom 3D printer and resin for 18 μm × 20 μm microfluidic flow channels. *Lab Chip* **2017**, *17*, 2899–2909. [CrossRef] [PubMed]
7. Cabrera-López, J.J.; García-Arrunátegui, M.F.; Neuta-Arciniegas, P.; Campo, O.; Velasco-Medina, J. PCB-3D-Printed, reliable and reusable wells for impedance spectroscopy of aqueous solutions. *J. Phys.* **2019**, *1272*, 012017. [CrossRef]
8. Autodesk. Tutorial: Mechanical to Electronics Workflows, Autodesk Fusion 360 Help Page. Available online: https://help.autodesk.com/view/fusion360/ENU/?guid=ECD-TUT-MECH-TO-ECAD-WORKFLOW-TOP-LEVEL (accessed on 22 February 2022).
9. Gong, H.; Beauchamp, M.; Perry, S.; Woolley, A.T.; Nordin, G.P. Optical Approach to Resin Formulation for 3D Printed Microfluidics. *RSC Adv.* **2015**, *5*, 106621–106632. [CrossRef] [PubMed]

MDPI

Article

Isothermal Recombinase Polymerase Amplification (RPA) of *E. coli* gDNA in Commercially Fabricated PCB-Based Microfluidic Platforms

Maria Georgoutsou-Spyridonos [1], Myrto Filippidou [1], Georgia D. Kaprou [1,†], Dimitrios C. Mastellos [2], Stavros Chatzandroulis [1] and Angeliki Tserepi [1,*]

[1] Institute of Nanoscience and Nanotechnology, NCSR-Demokritos, Patriarchou Gregoriou E' and 27 Neapoleos Str., Aghia Paraskevi, Attiki, 15341 Athens, Greece; m.georgoutsou-spyridonos@inn.demokritos.gr (M.G.-S.); m.filippidou@inn.demokritos.gr (M.F.); gdkaprou@gmail.com (G.D.K.); s.chatzandroulis@inn.demokritos.gr (S.C.)

[2] Institute of Nuclear & Radiological Sciences and Technology, Energy & Safety, NCSR-Demokritos, Patriarchou Gregoriou E' and 27 Neapoleos Str., Aghia Paraskevi, Attiki, 15341 Athens, Greece; mastellos@rrp.demokritos.gr

* Correspondence: a.tserepi@inn.demokritos.gr; Tel.: +30-210-650-3264

† Current address: Luxembourg Centre for Systems Biomedicine, University of Luxembourg, L-4367 Belvaux, Luxembourg.

Abstract: Printed circuit board (PCB) technology has been recently proposed as a convenient platform for seamlessly integrating electronics and microfluidics in the same substrate, thus facilitating the introduction of integrated and low-cost microfluidic devices to the market, thanks to the inherent upscaling potential of the PCB industry. Herein, a microfluidic chip, encompassing on PCB both a meandering microchannel and microheaters to accommodate recombinase polymerase amplification (RPA), is designed and commercially fabricated for the first time on PCB. The developed microchip is validated for RPA-based amplification of two *E. coli* target genes compared to a conventional thermocycler. The RPA performance of the PCB microchip was found to be well-comparable to that of a thermocycler yet with a remarkably lower power consumption (0.6 W). This microchip is intended for seamless integration with biosensors in the same PCB substrate for the development of a point-of-care (POC) molecular diagnostics platform.

Keywords: PCB technology; DNA amplification; RPA; microfluidics; microheaters; *E. coli*; molecular diagnostics

Citation: Georgoutsou-Spyridonos, M.; Filippidou, M.; Kaprou, G.D.; Mastellos, D.C.; Chatzandroulis, S.; Tserepi, A. Isothermal Recombinase Polymerase Amplification (RPA) of *E. coli* gDNA in Commercially Fabricated PCB-Based Microfluidic Platforms. *Micromachines* **2021**, *12*, 1387. https://doi.org/10.3390/mi12111387

Academic Editor: Nam-Trung Nguyen

Received: 10 October 2021
Accepted: 10 November 2021
Published: 12 November 2021

1. Introduction

The accommodation of conventional laboratory processes in microfluidic platforms fabricated using well-established microfabrication technology has drawn great attention and led to the development of the so-called lab-on-a-chip (LOC) devices. Typical advantages of microfluidic systems include the possibility to use very small quantities of expensive reagents and scarce samples, to perform high-resolution, precise, and sensitive detection, and to reduce the analysis time and cost [1]. Such systems are capable of performing a great variety of laboratory processes, such as sample purification and enrichment [2–4], reagent mixing [5,6], ultra-fast thermal cycling required in many biochemical reactions [7], as well as the detection of reaction products [8,9], which is of utmost importance. To achieve the above, LOC devices typically integrate microfluidic components, electrical driving circuits, and sensors into the same, usually hybrid, platform [8,10–12].

A drawback of such hybrid LOC platforms is that different substrate materials are used, and thus, different technologies need to be implemented for their fabrication, rendering the integration cumbersome and costly, thus hindering their commercial exploitation. Therefore, integrated LOC devices fabricated seamlessly on a single substrate are desirable and a fabrication technology amenable to mass production is sought to enhance the potential for commercialization of microfluidic-based diagnostic devices [8,13].

Nucleic acid analysis is an increasingly important tool for a wide variety of biochemical applications, such as molecular diagnostics, and as such, it is included as a step in many LOC devices and systems [14,15]. The uniqueness of nucleic acid sequences allows for the detection of biological agents with a high degree of specificity. However, since the concentration of nucleic acids in biologically derived analytes is low, their concentration must be amplified to be effectively used as a detection tool. This can be achieved by nucleic acid amplification (NAA) methods, such as the polymerase chain reaction (PCR), which is considered the gold standard in molecular diagnostics, as well as convenient isothermal amplification methods (including the helicase dependent amplification (HDA), the loop-mediated amplification (LAMP), and the recombinase polymerase amplification (RPA)), which are preferred for their simplified protocols and the elimination of thermocycling, which is accommodated in many LOC devices [16,17].

The implementation of (rigid or flexible) printed circuit board (PCB) substrates and the relevant technology for the realization of microdevices for NAA has been described in [18–22], proposing PCB technology as a convenient platform for integrating electronics and microfluidics, thus facilitating the introduction of such devices to the market, thanks to the inherent upscaling potential of the PCB industry and the well-established PCB technology around the world. PCB-based microdevices for static chamber [22] and continuous flow [21] PCR, as well as isothermal HDA-based DNA amplification [20] eliminating thermal cycling, have been described in the past. Amongst various isothermal NAA methods, RPA has attracted much attention due to its minimal sample preparation, increased sensitivity, specificity, robustness, and low cost, thus rendering it a perfect candidate for diagnostics in resource-limited settings [23]. Despite its advantages, an integrated PCB-based RPA microchip has not been yet reported.

In more detail, RPA is a NAA method carried out under isothermal conditions, thus not requiring thermocycling. Its main advantage is its low operational temperature near body temperature (37–42 °C), combined with its simplicity (minimal sample preparation, it can be carried out directly in serum, urine, as well as in the presence of known PCR inhibitors), sensitivity (down to 1–10 target DNA copies), and speed (5–20 min) [24,25]. The RPA method, first introduced by Piepenburg et al. [26], couples the isothermal recombinase-driven primer binding to the template DNA with the strand-displacement DNA synthesis. RPA is a compelling alternative to PCR, addressing the rapid detection of various pathogenic agents such as bacteria [23,27–31], viruses [32–34], parasites [35], and fungi [36]. In most of the previous works, the emphasis was on the development of POC diagnostic platforms for performing microbial analysis at the point-of-need; however, little attention was paid to the mass production of integrated chips, to allow for the massive deployment and adoption of microfluidics in the diagnostics market. Nevertheless, situations imposed by a pandemic such as COVID-19 provide additional motivation for the rapid development of new diagnostic microdevices [37] that are fast and massively fabricated by an established industry.

In this work, an RPA-on-PCB microdevice for performing DNA amplification was designed, commercially fabricated, and validated for performing DNA amplification of fragments of two genes of *E. coli*, which is a common Gram-negative bacterium that normally resides within the intestinal microbiota of humans. However, certain highly virulent *E. coli* strains can cause serious health conditions such as urinary tract infections (UTIs), respiratory illness, and pneumonia. While UTIs can result from both Gram-negative and Gram-positive bacterial expansion, the majority of UTI cases are attributed to *E. coli* strains [38]. Therefore, *E. coli* serves as a model pathogen for addressing UTI pathogenesis and developing relevant diagnostics. In this work, target gene-specific primers were designed and validated, while the RPA performance of the microdevice was compared to that of a conventional thermocycler. This microdevice is intended to be integrated in the same PCB substrate with biosensors for the development of a microdevice serving as a POC molecular diagnostics platform.

2. Materials and Methods

2.1. Design and Fabrication of an RPA-on-PCB Microchip

For the on-chip evaluation of the RPA amplification, a microchip was designed and fabricated on PCB, as the PCB technology allows for low-cost and standardized mass production, while the PCB substrate enables all the electrical connections required for the operation of the device. The chip was designed using an open source software, Kicad, whereas it was fabricated by a PCB manufacturing company, Eurocircuits (https://www.eurocircuits.com).

In particular, the chip dimensions were as follows: thickness: 1.55 mm, length: 65 mm, and width: 42 mm, and it consisted of a meandering microfluidic channel (occupying an area of 16 × 40 mm^2, Figure 1a), to minimize the chip footprint, on one side of the PCB substrate and a copper (Cu) microheater on the other side. The microchannel was patterned on a laminated photosensitive dry film (Figure 1a), and it had dimensions of 300 mm, 1 mm, and 100 μm for length, width and height, respectively, and a volume of 30 μL. Smaller volumes were also possible with the implementation of thinner photosensitive dry films. The microchannel was partly commercially fabricated on the solder mask layer, which was supplied together with the PCB from the PCB manufacturer (Eurocircuits, Kettenhausen, Germany), while for increased microfluidic channel height, the solder mask layer was combined in house with an ORDYL SY 300 film purchased from Resistechno. A cross-sectional view of the device is shown in Figure 1b. The microheater was implemented in the inner 18 μm-thick Cu layer of the PCB substrate (meandrous yellow tracks in Figure 1a) in the area below the microfluidic channel to allow for proper heating of the DNA amplification cocktail during the amplification, and it was isolated from the microfluidic channel by insulating layers. Furthermore, in order to optimize the temperature uniformity in the chip area where the microchannel was built [22], a solid Cu layer (magenta rectangle in Figure 1a) was formed on the inner 18 μm-thick Cu layer exactly under the microchannel. In order to provide for sample input and output, two through holes were opened at the beginning and end of the microchannel. Finally, the microchannel was sealed in house with a 100 μm polyolefin film (StarSeal from STARLAB) coated on one side with a silicone adhesive (PCR-compatible) in order to provide a strong adhesion even at elevated temperatures.

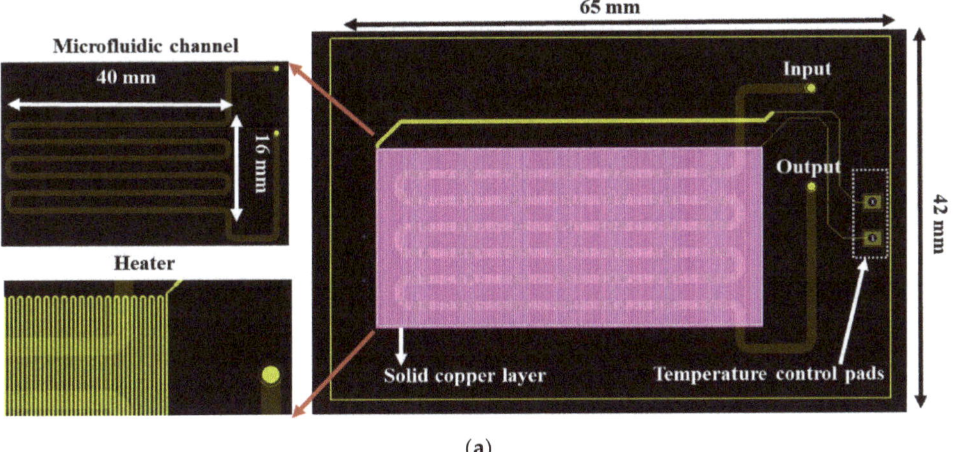

(a)

Figure 1. *Cont.*

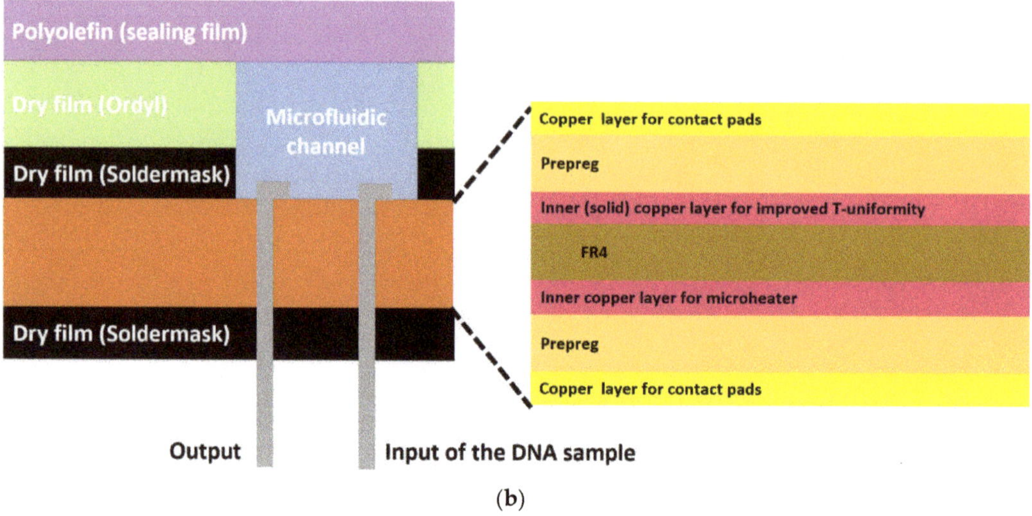

(**b**)

Figure 1. (**a**) RPA-on-PCB chip design for DNA amplification. The meandering microfluidic channel, the microheater with its electrical pads, as well as a solid copper layer beneath the microchannel for optimum temperature uniformity are depicted. (**b**) Schematic cross-sectional view of the RPA-on-PCB chip.

2.2. Temperature Control

Control of the microheater temperature was achieved using a temperature control unit providing the voltage needed across the resistive microheater, whilst a 100 mOhm resistor, placed in series with the microheater, was used to measure the current flowing through. Thus, the controller measures in real time the resistance of the microheater to derive its operating temperature through the temperature coefficient of resistance for copper (i.e., using the microheaters as temperature sensors) [21], and the resulting value was used in a proportional–integral (PI) feedback control loop to stabilize the temperature of the microheater at the desired set point.

2.3. Biological Assays

In this work, RPA was optimized and performed on a PCB-based microfluidic chip for the amplification of *E. coli* DNA. Toward this end, *E. coli* bacterial cultures were performed, followed by DNA extraction. Appropriate primers were also designed, and amplification methods were performed both in a conventional thermocycler and on chip. The assays are described in detail below.

2.3.1. Bacterial Culture

All experiments were performed with the strain of *Escherichia coli* TOP10. *E. coli* was grown overnight in Luria–Bertani (LB) medium (37 °C, 200 rpm); bacterial cells were inoculated in a fresh cultural medium (1:1000). Bacterial cultures were grown until mid-late exponential phase in liquid media. Cells were centrifuged and resuspended in 0.9% NaCl. Optical density was measured corresponding to different cell concentrations, 10^3–10^7 CFU/mL. To quantify the concentration of bacteria in saline solution (internal control), viable spread plate counts were determined by serial dilution plating on solid LB agar media.

2.3.2. DNA Extraction: Chemical and Thermal Lysis

Chemical Lysis: The genomic DNA (gDNA) from the bacterial cultures was extracted using enzymatic lysis buffer containing Proteinase K, RNase, Lysozyme, and SDS. The DNA was purified using phenol/chloroform solution and precipitated by ethanol. The concentration and the purity of the DNA samples were measured using a nanodrop spectrophotometer (Nanodrop 1000, ThermoFisher Scientific, Paisley, UK). Each DNA sample was standardized to 50 ng/µL and stored at −20 °C until use.

Thermal Lysis: First, 1 ml of *E. coli* culture in 0.9% NaCl was incubated for 10 min at 95 °C in thermoblock (Digital Dry Bath, Biorad). The sample was centrifuged (3000 rpm, RT °C 10 min), and the supernatant containing gDNA of *E. coli* was used as the cell lysate in the reactions.

2.3.3. Primer Design

According to the instructions for the RPA kit (TwistDX, Cambridge, UK, www.twistdx.co.uk/docs/default-source/RPA-assay-design) [39], RPA requires a different parameter design from that needed for PCR analysis (for example, a longer length of the primer, approximately 30–35 nucleotides). In addition, there are no melting temperature requirements for the design of RPA primers because hybridization and elongation are achieved by enzymes and are not induced by temperature [24]. Therefore, the primer design should consider multiple factors, including hairpin structure, mismatch, primer dimer, and amplification efficiency. In the present study, two target-specific primer pairs were designed for the amplification of a 210 bp fragment of the yBBW gene and a 176 bp DNA fragment of the malB gene. These primers were designed based on previously published primer sequences for the same *E. coli* genes and modifications including the extension of primer length. At the same time, the designed primers are also suitable for PCR amplification. The primer sequences used in this study are given in Table 1. All oligonucleotides were synthesized by Metabion International AG (Planegg, Germany) and purified by high-pressure liquid chromatography (HPLC). Oligonucleotides were delivered as dry, lyophilized powder, were dissolved in nuclease-free water at a concentration of 100 µM, and stored at 20 °C, in the dark.

Table 1. Primers designed in this study.

Name	Oligo Sequence	Length	GC%	Tm (°C)
FybbW	5′- TGA TTG GCA AAA TCT GGC CGG GAT TTT TAA CT-3′	31	38.7	61
RybbW	5′-GAA ATC GCC CAA ATC GCC ATA CCG CCG AAA AC-3′	32	53.1	66
FmalB	5′-GGA TAT TTC TGG TCC TGG TGC CGG-3′	24	58.3	62
RmalB	5′-TTT TCG ATG TGC GTT TAG CGC AGA-3′	24	45.8	60

2.3.4. Amplification Methods: PCR and RPA

For PCR, the KAPA2G Fast ReadyMix (KapaBiosystems, Wilmington, MA, USA) kit was used according to the supplier's instructions (https://www.kapabiosystems.com/product-applications/products/pcr-2/kapa2g-fast-pcr-kits/). KAPA 2G Fast DNA polymerase was mixed with 5 pmoles of each primer. Each 25 µL PCR reaction was supplemented with 1 µL of DNA template with a concentration of 1 ng/µL. DNA fragments encoding ybbW and malB genes were amplified, and each reaction contained 12.5 µL Buffer. Purified genomic *E. coli* TOP10 DNA (1 ng) was used as the template. The PCR protocol used in the thermocycler consisted of 3 steps of 10 s denaturation at 95 °C—10 s annealing at 60 °C (ybbW) and 55 °C (malB)—10 s extension at 72 °C, repeated 30 times. Subsequently, the PCR products were loaded on agarose gel (2%) stained with ethidium bromide and visualized with an ultraviolet (UV) visualizer.

RPA reactions were accomplished using commercially available RPA reagent kits provided in the TwistAmp Exo Kit, available from TwistDX Ltd. (Cambridge, UK). Each primer was used at a final concentration of 10 µM; the final volume was 25 µL. Purified genomic *Escherichia coli* TOP10 DNA (1 ng) was used as the template. RPA reactions were performed at 39 °C, for 30 min, in 50 µL volumes consisting of 29.5 µL of rehydration buffer, 2.4 µL of each primer (10 µM), a tube of lyophilized enzymes diluted in 16 µL dH_2O, and 1–2 ng of DNA template or 1 µL cell lysate. After thorough mixing, 2.5 µL of 280 mM MgOAc was added into the reaction system. Before the diluted enzymes were added to the reaction mixtures, they were irradiated with UV light in order to degrade and inactivate any DNA contamination, which could cause a false-positive amplification, as described previously [40]. In miniaturized amplification assays, RPA reactions were carried out for 10–30 min, in 12.5, 25, and 50 µL volumes. The experiments were executed in a conventional thermocycler (T-Personal 005-552, Biometra) and in a static PCB-based microdevice. RPA products were purified from enzymes and proteins using a NucleoSpin® Gel and PCR Clean-up kit (MACHEREY-NAGEL GmbH & Co, Düren, Germany). The purification procedure was necessary to visualize DNA bands in agarose gel electrophoresis. Alternatively, RPA products can be heat treated (5 min at 95 °C).

3. Results and Discussion

For the validation of the RPA-on-PCB microdevice, an isothermal protocol was optimized and followed for fast and efficient DNA amplification. Amplification reactions were carried out in the RPA microdevice and on a conventional thermocycler for comparison purposes. The results are presented and discussed in the following sections.

3.1. Selection of Primers for E. coli DNA Amplification

Escherichia coli are a common, large, and diverse group of bacteria found in the environment, food, and intestines of people and animals. Although most strains of *E. coli* are harmless, some can cause diarrhea, while others are responsible for 80 to 90% of the urinary tract infections. The RPA was selected as the amplification method because it is an isothermal one, thus avoiding thermocycling, while in addition, it can reduce analysis time much below 60 min. The *E. coli* strains have substantially diverse and multiple sequences. Alignment analysis of *E. coli* genomes reveals that the selection of universal and specific genetic targets in this bacterium is a great challenge. The target genes should be evolutionarily conserved to be found in all strains. In addition, it is essential to avoid the co-amplification of material extracted from other bacteria species or close related organisms. Therefore, the primer sequences should be carefully selected and designed to have 100% exclusivity and inclusivity.

A novel RPA assay was designed for the detection of two target genes ybbW and malB that, after in silico analysis, comply with the above requirements [41–43]. According to Walker et al. [41], the ybbW gene is part of the core genome (existing in >95% of all sequenced strains) of the *E. coli* offering great inclusivity (100%) and exclusivity (100%) within *E. coli*, whereas the malB gene is conserved across different *E. coli* lineages [44]. The ybbW gene sequence codes for a putative allantoin permease involved in the transport and metabolic conversion of allantoin, which is a metabolic intermediate that can serve as a source of nitrogen for bacterial cells under nutrient-limiting conditions (https://www.uniprot.org/uniprot/P75712). The malB gene sequence is derived from a large genomic region of *E coli* coding for a set of gene products (malB operon) involved in the transport, utilization, and metabolic turnover of maltose within bacterial cells. It is highly conserved among bacterial species. The region of the malB operon selected for *E coli*-specific primer design corresponds to a region that is discrete from other bacterial species and highly conserved among most *E coli* strains [45].

One set of novel primers was used for the successful amplification of each target gene. In addition, alignment studies of *E. coli* genes were completed using sequence information from the National Center for Biotechnology Information (NCBI) Genbank database. The primers were designed to be suitable for the RPA method and specific for amplification of the selected target genes (first, with a length 30–35 bp—applicable only for the ybbw, not for the malB primer set—and <45 bp, second, with a content in GC nucleotides >30% or <70%, and third, with amplification products 80–400 bp). Primer sequences were selected with the aid of Primer 3, and their selectivity for *E. coli* was determined using the Primer-BLAST algorithm. The primer analysis for dimers and hairpins was performed using the software offered by Integrated DNA Technologies (IDT, https://eu.idtdna.com/pages). The primers were tested in addition to RPA by the polymerase chain reaction (PCR) method in the thermal cycler. PCR was performed in a 25 μL reaction volume for DNA samples that comprised 1 μL purified DNA template (1 ng) and 10 μM primers. RPA was performed in a 50 μL reaction volume for purified 1 μL DNA or 1 μL cell lysate and 10 μM primers. The DNA amplification was verified via agarose gel electrophoresis. Figure 2 indicates that both primer pairs have excellent specificity for the target genes in both amplification methods, PCR and RPA. In more detail, both ybbW and malB genes are amplified with high efficiency and specificity in PCR, while ybbW is amplified more efficiently than malB in RPA (Image J analysis of the image in Figure 2B (i) indicates a four times higher fluorescence signal for ybbW). For this reason, from this point on in this work, the ybbW gene will be used in RPA for amplification of *E. coli* gDNA. In addition, images in Figure 2B (i) and (ii) indicate that RPA amplifies equivalently purified gDNA and gDNA lysed from *E. coli* cells; therefore, RPA robustness is demonstrated also in this work for requiring minimal sample preparation.

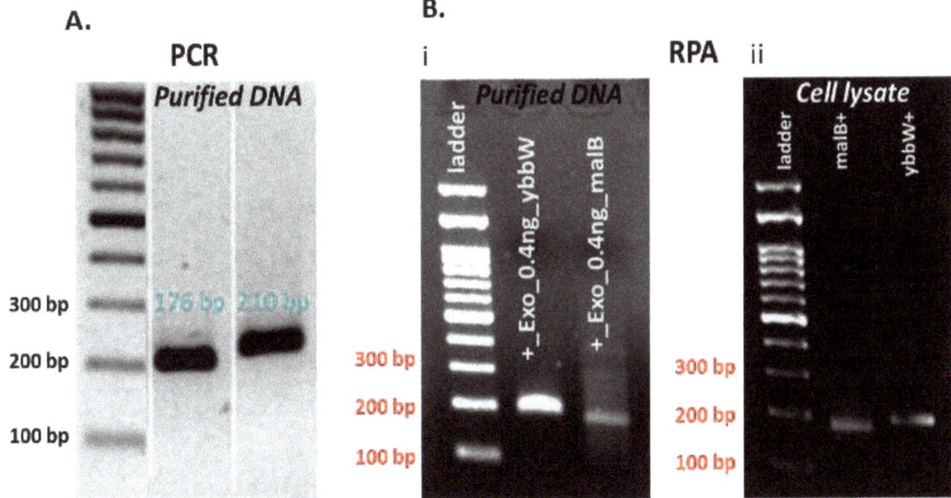

Figure 2. (**A**) Agarose gel electrophoresis (2%) of PCR products amplified by specific primer pairs for ybbW and malb genes with purified gDNA *E. coli* TOP10 (1 ng—2 × 10^5 copies DNA) template. (**B**) Agarose gel electrophoresis (2%) of RPA products amplified by target genes primer pairs with purified gDNA *E. coli* TOP10 (1 ng) and cell lysate (1 μL—2 × 10^7 copies DNA) (thermal lysis) template.

3.2. RPA Protocol and Optimization for On-Chip DNA Amplification

Point-of-care genetic diagnostics depend on the miniaturization of both the sample processing and the NAA devices. First, the heating at microscale level is supplied closer to the sample and is applied to a smaller thermal mass than in macroscale systems; this decreases the power consumption as well as the total reaction time [46], providing a faster time-to-result [16]. Furthermore, the miniaturization reduces the volume of the required amplification reagents, which is vital in amplification reactions in which costs are prohibitive to widespread use. Finally, miniaturization enables higher sensitivity and minimizes the risk of sample contamination. In isothermal enzymatic methods such as RPA, in vitro DNA synthesis occurs at a constant reaction temperature (39 °C), and hence, there is no need for an expensive thermal cycling instrument. However, several other factors are essential in performing RPA on a chip (with a static chamber), including the reaction volume and the amplification time.

First, control experiments for RPA miniaturization were performed on a thermal cycler. The DNA amplification was verified via agarose gel electrophoresis. Initially, samples were prepared as recommended by the kit manufacturer in a final volume of 50 μL/reaction. Then, they were divided into smaller fractions (1/2: 25 μL, $\frac{1}{4}$: 12.5 μL), as convenient sample volumes that can be safely loaded on PCB chips are 25 μL and 12.5 μL, compatible with fabricated microchannel volumes. Amplification reactions were performed for 30 min at 39 °C. For reducing the time-to-result, RPA reactions were also performed for 10 and 20 min. Figure 3 summarizes all RPA results from the cycler and indicates that RPA works with a satisfactory efficiency both in lower volumes (12.5 μL) and shorter time (10 min) than the kit manufacturer recommends, however with a lower amplification efficiency at shorter time (10 min).

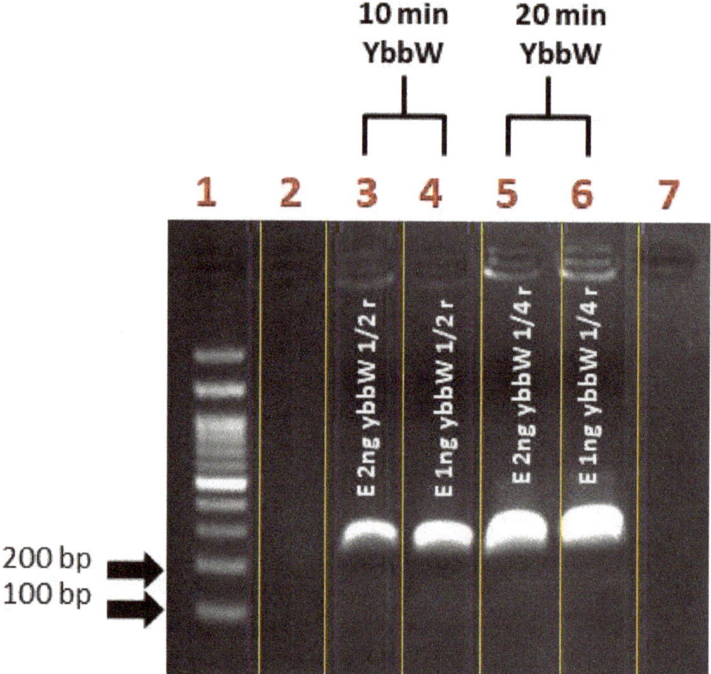

Figure 3. Agarose gel (2%) electrophoresis image of RPA reactions using ybbW primers and gDNA *E. coli* TOP10 (1–2 ng) as a template. The initial reaction was divided into different fractions before the amplification. The original TwistDx assay was performed for shorter time (Lane 3–4: 10 min and Lane 5–6: 20 min) and smaller sample volumes (Lane 3–4: 25 μL and Lane 5–6: 12.5 μL) than those recommended by the manufacturer. The performance of RPA was considered satisfactory.

3.3. *Characterization of the RPA-on-PCB Microdevice*

Figure 4a depicts the front (left) and the back (right) side of a fabricated RPA-on-PCB microdevice, while Figure 4b depicts the experimental set-up used for the evaluation of the RPA microdevice, comprising, in addition to the chip, the custom-made temperature controller unit and a laptop to facilitate user interfacing. The set-up was simplified by using a pipette (Figure 4a, right) for introducing (and collecting) the sample to the chip.

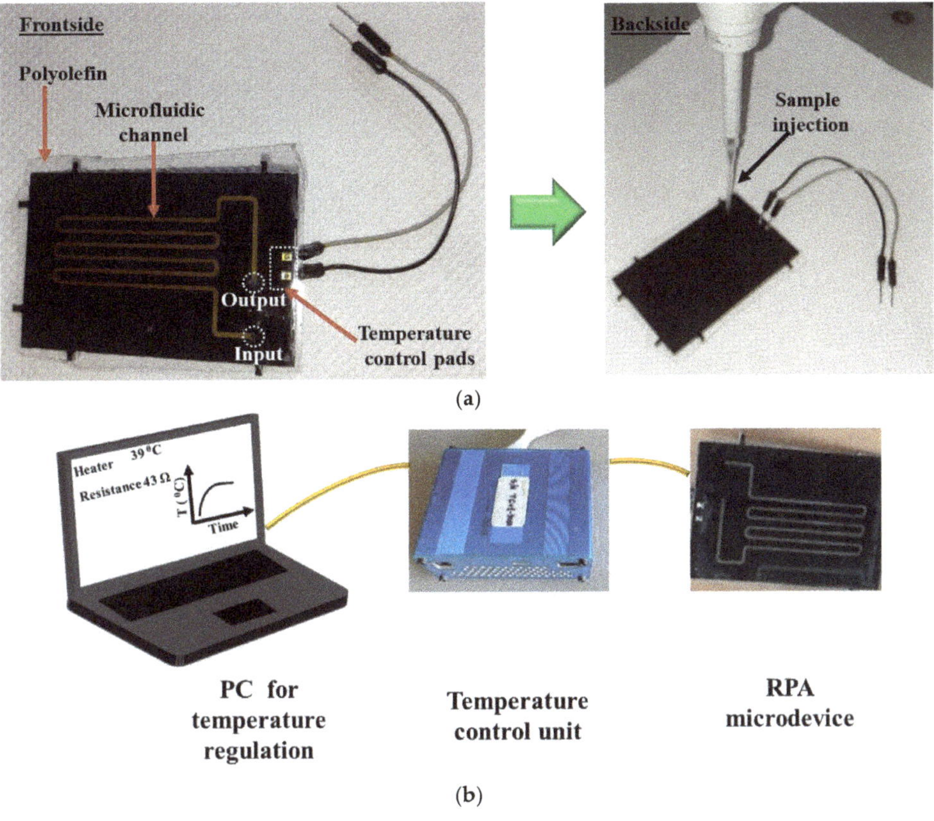

Figure 4. (**a**) Image of the frontside and backside of the RPA-on-PCB chip ready for use. The microfluidic channel and the sealing film (polyolefin) are depicted (left). Image of the backside of the device during the introduction of a RPA solution in the microchannel (right). (**b**) Schematic representation of the experimental set-up, comprising the RPA-on-PCB chip, the temperature control unit, and the PC with the user interface.

The embedded Cu microheater of the RPA-on-PCB chip was measured to have a resistance R_0 equal to 43 Ohm (measured at 25 °C), while the voltage, the current, and the power consumption were recorded during operation, to achieve and stabilize the temperature at the set point by means of the temperature controller. Figure 5a illustrates the temperature profile (red line) recorded by the temperature controller. The diagram indicates that after approximately 1.5 min, the temperature of the microheater reached the desirable set point (39 °C) starting from 28 °C and achieved stabilization at the set point within 5 min, with minimal fluctuations during the entire operation (30 min). In Figure 5b, the power consumption of the chip during operation is shown. After initial heating up from 28 °C, the microheater reached the set-point temperature (the current supplied was approximately 0.12 A), where the average power consumption was stabilized at 0.6 W. This power consumption is, as expected, smaller than that reported in continuous flow microPCR devices realized on PCB (2.7 W [21]) and far smaller than the power consumption of conventional thermocyclers (typically 500 W).

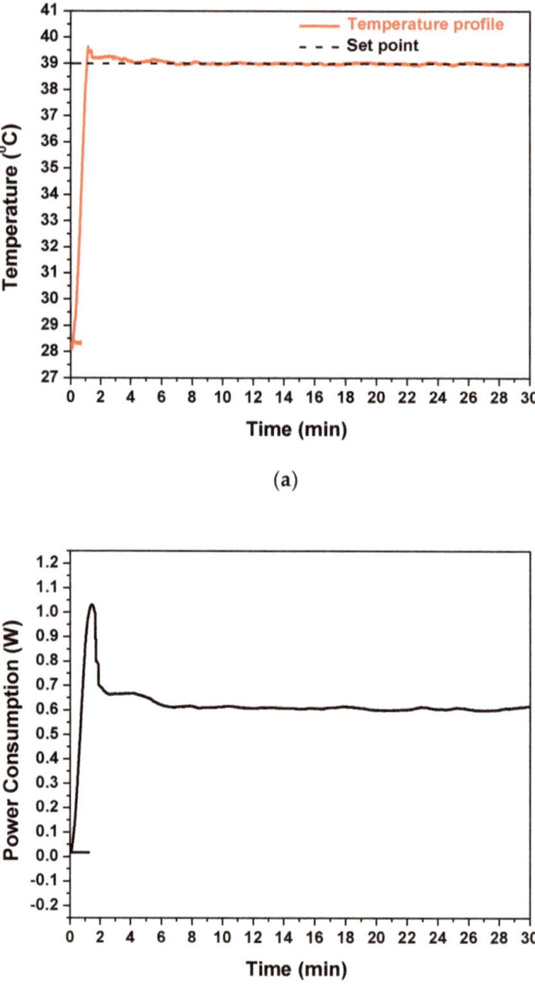

Figure 5. (**a**) The temperature profile of the embedded Cu microheater after heating up to set-point temperature and (**b**) its power consumption.

3.4. Validation of the RPA-on-PCB Microdevice

For performing on-chip RPA, the fabricated RPA-on-PCB microchip was connected to a custom-made temperature controller (Figure 4b) that senses the microheater's temperature, while on the other hand, it regulates the voltage across the microheater resistance to allow for precise control of the amplification temperature (39 °C). Before using each PCB microdevice for the first time, a washing step using ethanol was applied. For subsequent uses of the same device, an oxygen plasma step was applied to remove any biomolecule adsorbed on the microchannel surface. At this point, a 25 μL RPA solution containing *E. coli* DNA was introduced in the chamber using a micropipette (Figure 4a), and the temperature was maintained at 39 °C for 30 min, for performing DNA amplification. Simultaneously, a duplicate sample was amplified in the thermocycler as the positive control of the reaction. The amount of purified gDNA of *E. coli* TOP10 that was used as the reaction template was 2 ng. Negative control was also performed, where all reagents were added in the cocktail except for bacterial DNA. At the end of the experiments, the samples were collected by a micropipette and were cleaned up via thermal treatment. Amplification of the ybbW gene target was verified via gel electrophoresis (Figure 6), indicating a high amplification efficiency of the reaction performed on the PCB chip. In fact, analysis via Image J indicates slightly higher amplification on chip compared to that on the cycler. Thus, the results clearly demonstrate that the amplification of the ybbW gene target at 210 bp was successfully achieved in the developed RPA-on-PCB microdevice, with amplification efficiency well-comparable to that of a conventional thermocycler.

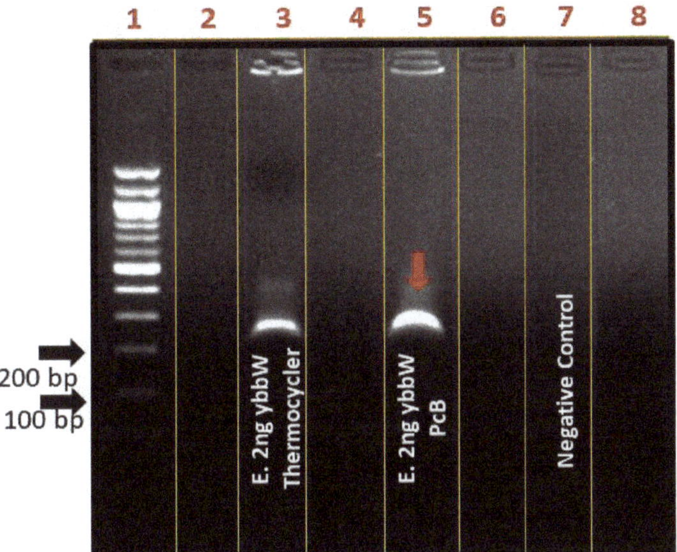

Figure 6. Agarose gel (2%) electrophoresis image of RPA reactions using ybbW primers and gDNA of *E. coli* TOP10 (2 ng) as a template. Lane 1: DNA ladder, Lane 3: positive control-ybbW amplified RPA product, RPA reaction on thermocycler, Lane 4: ybbW amplified RPA product, reaction on PCB chip, Lane 7: negative control-no gDNA, reaction on thermocycler.

The capability of PCB-based chips, similar to the present one, to perform PCR either in continuous flow or in static chamber microdevices has been demonstrated in the past [21,22]. The objective of this work was the demonstration of an RPA isothermal amplification as a simplified method not requiring thermocycling that is mostly appropriate for POC use.

4. Conclusions

In this article, we describe the development of a simple, low-cost microfluidic chip commercially fabricated for the first time on PCB, incorporating on the same PCB substrate a microchannel and resistive microheaters that are capable of performing RPA efficiently. The microchip was validated for achieving DNA amplification of two target genes of *E. coli*, which is a common bacterium potentially responsible for urinary tract infections, respiratory illness, and pneumonia. Specific primers were validated, while the RPA performance of the microchip was found to be well-comparable to that of a conventional thermocycler, yet with a remarkably lower power consumption. This microchip is intended in the near future to be seamlessly integrated with biosensors in the same PCB substrate for the development of a point-of-care molecular diagnostics platform.

Author Contributions: Conceptualization, A.T. and S.C.; methodology, G.D.K., M.G.-S. and D.C.M.; validation, G.D.K., M.G.-S. and M.F.; formal analysis, M.G.-S. and G.D.K.; writing—original draft preparation, M.G.-S., M.F. and A.T.; writing—review and editing, A.T., G.K. and D.C.M.; visualization, M.F. and M.G.-S.; supervision, A.T., S.C. and D.C.M.; project administration, A.T.; funding acquisition, A.T. and S.C. All authors have read and agreed to the published version of the manuscript.

Funding: This research was co-financed by the European Regional Development Fund of the European Union and Greek national funds through the Operational Program Competitiveness, Entrepreneurship and Innovation, under the call RESEARCH–CREATE–INNOVATE (project code: T1EDK-03565).

Acknowledgments: The authors wish to thank Anastasia Mpakali for providing expert assistance with molecular purification techniques and Efstratios Stratikos for allowing access to the Protein Chemistry Group Facilities at IPRETEA, NCSR "Demokritos". The authors also thank the Molecular Diagnostic laboratory of IPRETEA for allowing access to their gel imaging/documentation system. Furthermore, the authors would like to thank Panagiota Petrou for her scientific advice and for allowing access to the Immunochemistry Group facilities at IPRETEA.

Conflicts of Interest: The authors declare no conflict of interest.

References

1. Whitesides, G.M. The origins and the future of microfluidics. *Nature* **2006**, *442*, 368–373. [CrossRef]
2. Witek, M.A.; Llopis, S.D.; Wheatley, A.; McCarley, R.L.; Soper, S.A. Purification and preconcentration of genomic DNA from whole cell lysates using photoactivated polycarbonate (PPC) microfluidic chips. *Nucleic Acids Res.* **2006**, *34*, e74. [CrossRef]
3. Kastania, A.S.; Tsougeni, K.; Papadakis, G.; Gizeli, E.; Kokkoris, G.; Tserepi, A.; Gogolides, E. Plasma micro-nanotextured polymeric micromixer for DNA purification with high efficiency and dynamic range. *Anal. Chim. Acta* **2016**, *942*, 58–67. [CrossRef]
4. Tsougeni, K.; Papadakis, G.; Gianneli, M.; Grammoustianou, A.; Constantoudis, V.; Dupuy, B.; Petrou, P.S.; Kakabakos, S.E.; Tserepi, A.; Gizeli, E.; et al. Plasma nanotextured polymeric lab-on-a-chip for highly efficient bacteria capture and lysis. *Lab Chip* **2016**, *16*, 120–131. [CrossRef] [PubMed]
5. Papadopoulos, V.E.; Kefala, I.N.; Kaprou, G.; Kokkoris, G.; Moschou, D.; Papadakis, G.; Gizeli, E.; Tserepi, A. A passive micromixer for enzymatic digestion of DNA. *Microelectron. Eng.* **2014**, *124*, 42–46. [CrossRef]
6. Kefala, I.N.; Papadopoulos, V.E.; Karpou, G.; Kokkoris, G.; Papadakis, G.; Tserepi, A. A labyrinth split and merge micromixer for bioanalytical applications. *Microfluid. Nanofluid.* **2015**, *19*, 1047–1059. [CrossRef]
7. Lee, D.-S.; Park, S.H.; Yang, H.; Chung, K.-H.; Yoon, T.H.; Kim, S.-J.; Kim, K.; Kim, Y.T. Bulk-micromachined submicroliter-volume PCR chip with very rapid thermal response and low power consumption. *Lab Chip* **2004**, *4*, 401–407. [CrossRef] [PubMed]
8. Lei, K.F. Microfluidic Systems for Diagnostic Applications: A Review. *J. Lab. Autom.* **2012**, *17*, 330–347. [CrossRef]
9. Tsougeni, K.; Kaprou, G.; Loukas, C.M.; Papadakis, G.; Hamiot, A.; Eck, M.; Rabus, D.; Kokkoris, G.; Chatzandroulis, S.; Papadopoulos, V.; et al. Lab-on-Chip platform and protocol for rapid foodborne pathogen detection comprising on-chip cell capture, lysis, DNA amplification and surface-acoustic-wave detection. *Sens. Actuators B Chem.* **2020**, *320*, 128345. [CrossRef]
10. Schumacher, S.; Nestler, J.; Otto, T.; Wegener, M.; Ehrentreich-Förster, E.; Michel, D.; Wunderlich, K.; Palzer, S.; Sohn, K.; Weber, A.; et al. Highly-integrated lab-on-chip system for point-of-care multiparameter analysis. *Lab Chip* **2012**, *12*, 464–473. [CrossRef]
11. Papadakis, G.; Murasova, P.; Hamiot, A.; Tsougeni, K.; Kaprou, G.; Eck, M.; Rabus, D.; Bilkova, Z.; Dupuy, B.; Jobst, G.; et al. Micro-nano-bio acoustic system for the detection of foodborne pathogens in real samples. *Biosens. Bioelectron.* **2018**, *111*, 52–58. [CrossRef]
12. Tsougeni, K.; Kastania, A.S.; Kaprou, G.D.; Eck, M.; Jobst, G.; Petrou, P.S.; Kakabakos, S.E.; Mastellos, D.; Gogolides, E.; Tserepi, A. A modular integrated lab-on-a-chip platform for fast and highly efficient sample preparation for foodborne pathogen screening. *Sens. Actuators B Chem.* **2019**, *288*, 171–179. [CrossRef]
13. Moschou, D.; Tserepi, A. The lab-on-PCB approach: Tackling the μTAS commercial upscaling bottleneck. *Lab Chip* **2017**, *17*, 1388–1405. [CrossRef]
14. Zhang, Y.; Ozdemir, P. Microfluidic DNA amplification—A review. *Anal. Chim. Acta* **2009**, *638*, 115–125. [CrossRef] [PubMed]
15. Gorgannezhad, L.; Stratton, H.; Nguyen, N.-T. Microfluidic-Based Nucleic Acid Amplification Systems in Microbiology. *Micromachines* **2019**, *10*, 408. [CrossRef] [PubMed]
16. Ahmad, F.; Hashsham, S. Miniaturized nucleic acid amplification systems for rapid and point-of-care diagnostics: A review. *Anal. Chim. Acta* **2012**, *733*, 1–15. [CrossRef] [PubMed]
17. Obande, G.A.; Banga Singh, K.K. Current and Future Perspectives on Isothermal Nucleic Acid Amplification Technologies for Diagnosing Infections. *Infect. Drug Resist.* **2020**, *13*, 455–483. [CrossRef] [PubMed]
18. Moschou, D.; Vourdas, N.; Filippidou, M.; Tsouti, V.; Kokkoris, G.; Tsekenis, G.; Zergioti, I.; Chatzandroulis, S.; Tserepi, A. *Integrated Biochip for PCR-Based DNA Amplification and Detection on Capacitive Biosensors*; SPIE: Cergy Pontoise Cedex, France, 2013; Volume 8765.

19. Aditya, R. Microfabricated Tools and Engineering Methods for Sensing Bioanalytes. Ph.D. Thesis, California Institute of Technology, Pasadena, CA, USA, 2014.
20. Kaprou, G.D.; Papadakis, G.; Papageorgiou, D.P.; Kokkoris, G.; Papadopoulos, V.; Kefala, I.; Gizeli, E.; Tserepi, A. Miniaturized devices for isothermal DNA amplification addressing DNA diagnostics. *Microsyst. Technol.* **2016**, *22*, 1529–1534. [CrossRef]
21. Kaprou, G.D.; Papadopoulos, V.; Papageorgiou, D.P.; Kefala, I.; Papadakis, G.; Gizeli, E.; Chatzandroulis, S.; Kokkoris, G.; Tserepi, A. Ultrafast, low-power, PCB manufacturable, continuous-flow microdevice for DNA amplification. *Anal. Bioanal. Chem.* **2019**, *411*, 5297–5307. [CrossRef]
22. Kaprou, G.D.; Papadopoulos, V.; Loukas, C.-M.; Kokkoris, G.; Tserepi, A. Towards PCB-Based Miniaturized Thermocyclers for DNA Amplification. *Micromachines* **2020**, *11*, 258. [CrossRef]
23. Ereku, L.T.; Mackay, R.E.; Craw, P.; Naveenathayalan, A.; Stead, T.; Branavan, M.; Balachandran, W. RPA using a multiplexed cartridge for low cost point of care diagnostics in the field. *Anal. Biochem.* **2018**, *547*, 84–88. [CrossRef]
24. Daher, R.K.; Stewart, G.; Boissinot, M.; Bergeron, M.G. Recombinase Polymerase Amplification for Diagnostic Applications. *Clin. Chem.* **2016**, *62*, 947–958. [CrossRef] [PubMed]
25. Lobato, I.M.; O'Sullivan, C.K. Recombinase polymerase amplification: Basics, applications and recent advances. *Trends Analyt. Chem.* **2018**, *98*, 19–35. [CrossRef] [PubMed]
26. Piepenburg, O.; Williams, C.H.; Stemple, D.L.; Armes, N.A. DNA Detection Using Recombination Proteins. *PLoS Biol.* **2006**, *4*, e204. [CrossRef]
27. Dao, T.N.T.; Lee, E.Y.; Koo, B.; Jin, C.E.; Lee, T.Y.; Shin, Y. A microfluidic enrichment platform with a recombinase polymerase amplification sensor for pathogen diagnosis. *Anal. Biochem.* **2018**, *544*, 87–92. [CrossRef]
28. Lutz, S.; Weber, P.; Focke, M.; Faltin, B.; Hoffmann, J.; Müller, C.; Mark, D.; Roth, G.; Munday, P.; Armes, N.; et al. Microfluidic lab-on-a-foil for nucleic acid analysis based on isothermal recombinase polymerase amplification (RPA). *Lab Chip* **2010**, *10*, 887–893. [CrossRef] [PubMed]
29. Shen, F.; Davydova, E.K.; Du, W.; Kreutz, J.E.; Piepenburg, O.; Ismagilov, R.F. Digital isothermal quantification of nucleic acids via simultaneous chemical initiation of recombinase polymerase amplification reactions on SlipChip. *Anal. Chem.* **2011**, *83*, 3533–3540. [CrossRef]
30. Renner, L.D.; Zan, J.; Hu, L.I.; Martinez, M.; Resto, P.J.; Siegel, A.C.; Torres, C.; Hall, S.B.; Slezak, T.R.; Nguyen, T.H.; et al. Detection of ESKAPE Bacterial Pathogens at the Point of Care Using Isothermal DNA-Based Assays in a Portable Degas-Actuated Microfluidic Diagnostic Assay Platform. *Appl. Environ. Microbiol.* **2017**, *83*, e02449-16. [CrossRef]
31. Yeh, E.-C.; Fu, C.-C.; Hu, L.; Thakur, R.; Feng, J.; Lee, L.P. Self-powered integrated microfluidic point-of-care low-cost enabling (SIMPLE) chip. *Sci. Adv.* **2017**, *3*, e1501645. [CrossRef]
32. Rohrman, B.A.; Richards-Kortum, R.R. A paper and plastic device for performing recombinase polymerase amplification of HIV DNA. *Lab Chip* **2012**, *12*, 3082–3088. [CrossRef]
33. Magro, L.; Escadafal, C.; Garneret, P.; Jacquelin, B.; Kwasiborski, A.; Manuguerra, J.-C.; Monti, F.; Sakuntabhai, A.; Vanhomwegen, J.; Lafaye, P.; et al. Paper microfluidics for nucleic acid amplification testing (NAAT) of infectious diseases. *Lab Chip* **2017**, *17*, 2347–2371. [CrossRef]
34. Liu, D.; Shen, H.; Zhang, Y.; Shen, D.; Zhu, M.; Song, Y.; Zhu, Z.; Yang, C. A microfluidic-integrated lateral flow recombinase polymerase amplification (MI-IF-RPA) assay for rapid COVID-19 detection. *Lab Chip* **2021**, *21*, 2019–2026. [CrossRef]
35. Cordray, M.S.; Richards-Kortum, R.R. A paper and plastic device for the combined isothermal amplification and lateral flow detection of Plasmodium DNA. *Malar. J.* **2015**, *14*, 472. [CrossRef]
36. Lau, H.Y.; Wang, Y.; Wee, E.J.H.; Botella, J.R.; Trau, M. Field Demonstration of a Multiplexed Point-of-Care Diagnostic Platform for Plant Pathogens. *Anal. Chem.* **2016**, *88*, 8074–8081. [CrossRef]
37. Berkenbrock, J.A.; Grecco-Machado, R.; Achenbach, S. Microfluidic devices for the detection of viruses: Aspects of emergency fabrication during the COVID-19 pandemic and other outbreaks. *Proc. Math. Phys. Eng. Sci.* **2020**, *476*, 20200398. [CrossRef]
38. Foxman, B.; Brown, P. Epidemiology of urinary tract infections: Transmission and risk factors, incidence, and costs. *Infect. Dis. Clin. N. Am.* **2003**, *17*, 227–241. [CrossRef]
39. TwistDx™. *TwistAmp®DNA Amplification Kits Assay Design Manual*; TwistDx™: Cambridge, UK, 2018; p. 32.
40. McQuillan, J.S.; Wilson, M.W. Recombinase polymerase amplification for fast, selective, DNA-based detection of faecal indicator Escherichia coli. *Lett. Appl. Microbiol.* **2021**, *72*, 382–389. [CrossRef] [PubMed]
41. Walker, D.I.; McQuillan, J.; Taiwo, M.; Parks, R.; Stenton, C.A.; Morgan, H.; Mowlem, M.C.; Lees, D.N. A highly specific Escherichia coli qPCR and its comparison with existing methods for environmental waters. *Water Res.* **2017**, *126*, 101–110. [CrossRef] [PubMed]
42. Connelly, J.T.; Rolland, J.P.; Whitesides, G.M. "Paper Machine" for Molecular Diagnostics. *Anal. Chem.* **2015**, *87*, 7595–7601. [CrossRef]
43. Hill, J.; Beriwal, S.; Chandra, I.; Paul, V.K.; Kapil, A.; Singh, T.; Wadowsky, R.M.; Singh, V.; Goyal, A.; Jahnukainen, T.; et al. Loop-mediated isothermal amplification assay for rapid detection of common strains of Escherichia coli. *J. Clin. Microbiol.* **2008**, *46*, 2800–2804. [CrossRef]
44. Stratakos, A.C.; Linton, M.; Millington, S.; Grant, I.R. A loop-mediated isothermal amplification method for rapid direct detection and differentiation of nonpathogenic and verocytotoxigenic Escherichia coli in beef and bovine faeces. *J. Appl. Microbiol.* **2017**, *122*, 817–828. [CrossRef] [PubMed]

45. Dahl, M.K.; Francoz, E.; Saurin, W.; Boos, W.; Manson, M.D.; Hofnung, M. Comparison of sequences from the malB regions of Salmonella typhimurium and Enterobacter aerogenes with Escherichia coli K12: A potential new regulatory site in the interoperonic region. *Mol. Gen. Genet. MGG* **1989**, *218*, 199–207. [CrossRef] [PubMed]
46. Papadopoulos, V.E.; Kokkoris, G.; Kefala, I.N.; Tserepi, A. Comparison of continuous-flow and static-chamber μPCR devices through a computational study: The potential of flexible polymeric substrates. *Microfluid. Nanofluid.* **2015**, *19*, 867–882. [CrossRef]

 *micromachines* 

Article

Utilising Commercially Fabricated Printed Circuit Boards as an Electrochemical Biosensing Platform

Uroš Zupančič, Joshua Rainbow, Pedro Estrela and Despina Moschou *

Centre for Biosensors, Bioelectronics and Biodevices (C3Bio), Department of Electronic & Electrical Engineering, University of Bath, Claverton Down, Bath BA2 7AY, UK; uz206@bath.ac.uk (U.Z.); jr993@bath.ac.uk (J.R.); p.estrela@bath.ac.uk (P.E.)

* Correspondence: D.Moschou@bath.ac.uk; Tel.: +44-(0)-1225-383245

Abstract: Printed circuit boards (PCBs) offer a promising platform for the development of electronics-assisted biomedical diagnostic sensors and microsystems. The long-standing industrial basis offers distinctive advantages for cost-effective, reproducible, and easily integrated sample-in-answer-out diagnostic microsystems. Nonetheless, the commercial techniques used in the fabrication of PCBs produce various contaminants potentially degrading severely their stability and repeatability in electrochemical sensing applications. Herein, we analyse for the first time such critical technological considerations, allowing the exploitation of commercial PCB platforms as reliable electrochemical sensing platforms. The presented electrochemical and physical characterisation data reveal clear evidence of both organic and inorganic sensing electrode surface contaminants, which can be removed using various pre-cleaning techniques. We demonstrate that, following such pre-treatment rules, PCB-based electrodes can be reliably fabricated for sensitive electrochemical biosensors. Herein, we demonstrate the applicability of the methodology both for labelled protein (procalcitonin) and label-free nucleic acid (*E. coli*-specific DNA) biomarker quantification, with observed limits of detection (LoD) of 2 pM and 110 pM, respectively. The proposed optimisation of surface pre-treatment is critical in the development of robust and sensitive PCB-based electrochemical sensors for both clinical and environmental diagnostics and monitoring applications.

Keywords: printed circuit boards; electrochemical biosensors; Lab-on-PCB; electrode pre-treatment

Citation: Zupančič, U.; Rainbow, J.; Estrela, P.; Moschou, D. Utilising Commercially Fabricated Printed Circuit Boards as an Electrochemical Biosensing Platform. *Micromachines* **2021**, *12*, 793. https://doi.org/10.3390/mi12070793

Academic Editor: Francisco Perdigones

Received: 2 June 2021
Accepted: 1 July 2021
Published: 3 July 2021

1. Introduction

While the first idea for printed circuit boards originated in the early 1930s, the industry has come a significant way and expanded into almost every sector within the field of electronics, becoming a ubiquitous part of our everyday lives. Printed circuit boards (PCBs) have inarguably opened the door to the huge technological expansion and development of our times. Thus, there has been constant pressure for technological advancement, pushing for improved PCB performance, cost-effectiveness, and miniaturisation, over the last twenty-five years [1]. One application area facilitated by this advancement in high-specification commercial PCB technology is the sensors and diagnostics field, including the development of biosensors for the detection of biomolecules in clinical and environmental applications [2]. While first stipulated in the 1990s as a concept for integrated microchips for biomolecular detection [3], PCBs have recently re-emerged as a promising platform for the development of fully integrated and electronics-enabled Lab-on-Chip (LoC) platforms. To date, a large proportion of the conducted research has been focusing on microfluidic component and biosensor prototyping by complementary metal oxide semiconductor (CMOS) and polymer material platforms [4]. However, the development of LoC devices addressing demanding biomedical applications requires the integration of not only electronic sensor components but also sample preparation microfluidic components, heating, and filtration elements as well as fluid actuators and a user-friendly interface with the capability of data transmission and storage. The inherent characteristic of PCBs to be easily integrated with electronics in cm-scale devices in an unscalable and modular fashion has proven to be a distinct advantage, fuelling its recent re-emergence as Lab-on-PCB technology.

The concept of Lab-on-PCB devices combines high sensor performance (sensitivity and selectivity), cost-effectiveness, and the ability to be easily integrated [5]. Lab-on-Chip devices should meet the REASSURED criteria for point-of-care diagnostics if they are to be used for clinical purposes. This requires the sensor to be capable of real-time analysis, with ease of sample collection, affordability, sensitivity, specificity, user-friendliness, speed, equipment-free, and to be capable of being delivered to the patient within their location [5–7]. PCBs can realise real-time analysis through the interfacing of sensing electrodes and miniaturised electronics for electronic data acquisition, transmission, and storage.

Despite these unique advantages, new technological hurdles have emerged while developing commercial Lab-on-PCB platforms, arising from the fact that the industrial PCB fabrication process was not initially designed for use in the context of diagnostics. Thus, there is a major question that needs to be addressed when utilising these devices for highly sensitive and reliable electrochemical sensors. The fabrication process of commercial PCBs is highly industrial and comes with a plethora of contaminants that must be removed before use. These contaminants exist in both organic and inorganic forms and can cause issues including low sensitivity and signal interference. Issues can also arise in low fidelity with certain surface chemistry functionalisation techniques if not properly pre-cleaned before use.

This paper aims to characterise both the electrochemical and physical properties of PCB gold electrodes fabricated using a standard, industrial process (hard gold plating). We take a specific interest in the post-fabrication contamination and techniques for minimising it prior to constructing an electrochemical biosensor, as well as the sensing electrode surface roughness parameter. Finally, the paper proves the validity of these approaches, demonstrating two examples of sensitive PCB-based electrochemical biosensors for the detection of clinically relevant protein and nucleic acid biomarkers.

2. Materials and Methods

Hydrogen peroxide, ammonium hydroxide, potassium hydroxide, potassium ferrocyanide and ferricyanide, phosphate-buffered saline tablets, potassium chloride, and copper etchant (CE-100) were purchased from Sigma-Aldrich (Gillingham, UK), while 1 M sulphuric acid was purchased from Thermo Fisher Scientific (Loughborough, UK). The Ag/AgCl (KCl) reference electrode was purchased from BASi (West Lafayette, IN, USA) and the platinum wire used as the counter-electrode was obtained from ALS (Tokyo, Japan). Milli-Q water was obtained using the Millipore Direct-Q 5 UV Water Purification System and deionised (DI) water. Oxygen plasma Zepto System (Diener electronic, Ebhausen, Germany) was used to perform PCB plasma cleaning.

PCBs were designed using Altium Designer 18 software and fabricated by Lyncolec (Dorset, UK). In short, 1.6 mm FR-4 covered with 1 oz copper was patterned and plated with hard gold by electrodeposition of nickel (3–5 μm) and gold (1 μm) and outlined by the solder mask to make PCB electrodes with 1 mm diameter. PCB boards used in the study are shown in Figure S1. Pictures were taken using a Huawei P10 smartphone.

2.1. Surface Roughness Characterisation

AFM analysis was performed using Digital Instruments Nanoscope IIIA, and Gwyddion software was used for image analysis and profile extraction. SPR chips (Reichert Technologies, Buffalo, NY, USA) coated with a thin gold layer through vacuum deposition were used as planar electrodes for surface characterisation. For electrochemical characterisation, the chip was first cleaned using piranha solution (9 mL of 99.9% sulphuric acid, mixed with 3 mL of 30% hydrogen peroxide, for 5 min) before washing in MQ water. Electrodes were outlined by double-sided adhesive (300LSE, 3M, Bracknell, UK), where a 2 mm diameter circular hole was cut using a puncher. Then, 25 μL of 50 mM H_2SO_4 was deposited on the outlined electrode and contacted with the Ag/AgCl (KCl) reference electrode and platinum wire counter-electrode. Sulphuric acid cycling was performed using an Ag/AgCl (KCl) reference electrode and platinum wire counter-electrode between −0.2 and +1.5 V at 200 mV/s. Roughness factor was calculated by integration of the gold-oxide reduction peak as described previously [8] by calculation of electrochemical surface area (*ESA*):

$$Q = \frac{1}{v_r}\int i{\cdot}V\prime dV\prime$$

where Q is the charge, v_r is the scan rate, and the integral $\int i{\cdot}V\prime dV\prime$ is the area of the gold oxide reduction peak. *ESA* can then be determined by:

$$ESA = \frac{Q}{390{\cdot}10^{-6}}$$

Furthermore, the surface roughness factor (R_f) can be calculated:

$$R_f = \frac{ESA}{A}$$

PCBs were cleaned using a LT SC-1 cleaning procedure before AFM analysis, followed by cycling in H_2SO_4 as described above to acquire surface roughness factors.

2.2. PCB Cleaning and Electrochemical Analytical Techniques

PCB boards represented in Figure S1 were used in the cleaning optimisation study. The cleaning procedures are described in Table 1 and Table S1.

Table 1. Cleaning procedures used in the study.

Cleaning Method	Procedure
Oxygen plasma treatment	3, 5, or 10 min at 100 W and 0.2 mbar (Diener Zepto System, Diener electronic, Ebhausen, Germany).
KOH/H_2O_2 treatment	Immersion in a solution of 30% H_2O_2 and 50 mM KOH for 10 min.
LT SC-1 clean	Step 1: Immersion in a solution of 30% NH_4OH, 30% H_2O_2, and MQ water in a ratio of 1:1:5 for 15 min. Step 2: Immersion in >99% acetone solution for 5 min. Step 3: Immersion in >99% IPA solution for 5 min. Step 4: Immersion in MQ water for 5 min.

Oxidation and reduction of potassium ferri-/ferrocyanide was evaluated by cyclic voltammetry (CV) in 5 mM ferri-/ferrocyanide couple in PBS with 1 M KCl, scanning between −0.2 and 0.7 V vs. Ag/AgCl (KCl) at 100 mV/s scan rate. The current density was obtained by integration of oxidation peak and dividing the peak height with the geometrical area of the electrode. Charge transfer resistance was obtained by EIS scan in the abovementioned solution by scanning from 100,000 Hz to 1 Hz at 10 mV amplitude at the DC bias of the formal potential observed in a CV (e.g., 0.245 V vs. Ag/AgCl (KCl)). The obtained plot was fitted with Randles equivalent circuit to extract the Rct value. For evaluation of the impurity peaks, CV was performed in PBS from −0.3 to 0.8 V vs. Ag/AgCl (KCl) at 1 V/s scan rate and peaks at approximately 0.35 V vs. Ag/AgCl (KCl) were evaluated.

2.3. Electrochemical Sensor for E. coli DNA Detection

PCB electrodes were cleaned utilising the low-temperature standard clean-1 (LT SC-1) method as detailed above. Following this cleaning, electrodes were passively functionalised using an optimised 1:15 molar ratio of 1 μM thiolated single stranded peptide nucleic acid (ssPNA) and 1 μM 6-Mercapto-1-Hexanol (MCH) in 50% dimethyl sulfoxide (DMSO) diluted in Milli-Q (18.2 MΩ.cm at 25 °C). This was performed at 4 °C for approximately

16 h to form a self-assembled monolayer as described previously [5]. After functionalisation, electrodes were rinsed with Milli-Q and backfilled with 1 mM MCH diluted in 10 mM phosphate buffer (PB) for 50 min at room temperature. Electrodes were then rinsed with Milli-Q and incubated in measurement buffer (10 mM PB containing 4 mM $[Fe(CN)_6]^{3/4-}$) for 2 h to stabilise the self-assembly monolayer (SAM). The target ssDNA sequence was diluted in 10 mM PB and heated to 95 °C for 5 min prior to incubation. A sample volume of 10 µL was incubated on each electrode for 30 min before measuring signal change upon hybridisation using electrochemical impedance spectroscopy (EIS). The measurements were taken using the on-board three-electrode setup with a gold PCB quasi-reference electrode. Thus, the EIS spectra were scanned between 100,000 and 0.1 Hz with no DC bias and potential amplitude of 0.01 V versus open circuit potential (OCP). The calibration curve data were fit with a non-linear curve fitting. Change from the blank was subtracted before averaging data and fitting with Randle's equivalent circuit to obtain charge transfer resistance (ΔR_{ct}).

2.4. Electrochemical Detection of PCT Protein

ELISA was performed using PCT antibody pair (Abcam, Cambridge, UK, part no. ab222276) with Nunc MaxiSorp^{TM} high protein-binding 96-well ELISA plates by diluting the capturing antibody to 2 µg/mL in PBS and coating the plate for 2 h (100 µL/well) while shaking (200 rpm). Blocking was performed using 1% BSA in PBS for 1 h while shaking. PCT serial dilution was performed according to the manufacturer's protocol and the detection antibody was incubated with the target at the final concentration of 0.25 µg/mL for 1.5 h while shaking. Poly-HRP Streptavidin (Thermo Fisher Scientific, Loughborough, UK, part no. N200) was diluted 1:5000 in 1% BSA solution and incubated for 5 min while shaking. Color Reagent A and B containing 3,3′,5,5′-Tetramethylbenzidine (TMB) were mixed in a ratio of 1:1 (RnD Systems, Abingdon, UK, part no. DY007) and incubated on the plate for 2 min before being pipetted into a separate well to stop the reaction. Optical evaluation was performed using micro-volume spectrophotometer Genova Nano (Jenway, Staffordshire, UK) with 0.2 mm path length by measuring absorbance at 650 nm. The data were fitted with exponential function in Origin 9.1 software. Electrochemical measurements were performed by drop-casting the TMB solution on the PCB electrode surface and performing chronoamperometry at 0.1 DC bias vs. gold PCB quasi-reference electrode. The data point at 30 sec was used for quantification; three separate electrodes were used for obtaining the data. The change in blank was subtracted from the measurements before averaging. Hill fit was used to fit the data.

3. Results and Discussion

3.1. Removal of Surface Impurities in Commercially Manufactured PCB Electrodes

To use PCB electrodes as electrochemical sensors, the electrochemical behaviour of the bare electrodes must be repeatable and reproducible. Due to the low-cleanliness environment of the PCB manufacturing facilities, a level of impurities is present in the electrode surface after the production [5,9,10].

A pre-treatment step is therefore required to obtain a clean electroactive surface. Non-treated PCB electrodes revealed small currents in ferri-/ferrocyanide solution during the first scan and an increase in current density in the second scan to 4.6 µA/mm^2, which remained stable in the following scans (Figure 1a,b). This indicates the presence of an insulating layer on the PCB electrodes, which is removed upon the potential increase, allowing redox reactions to occur.

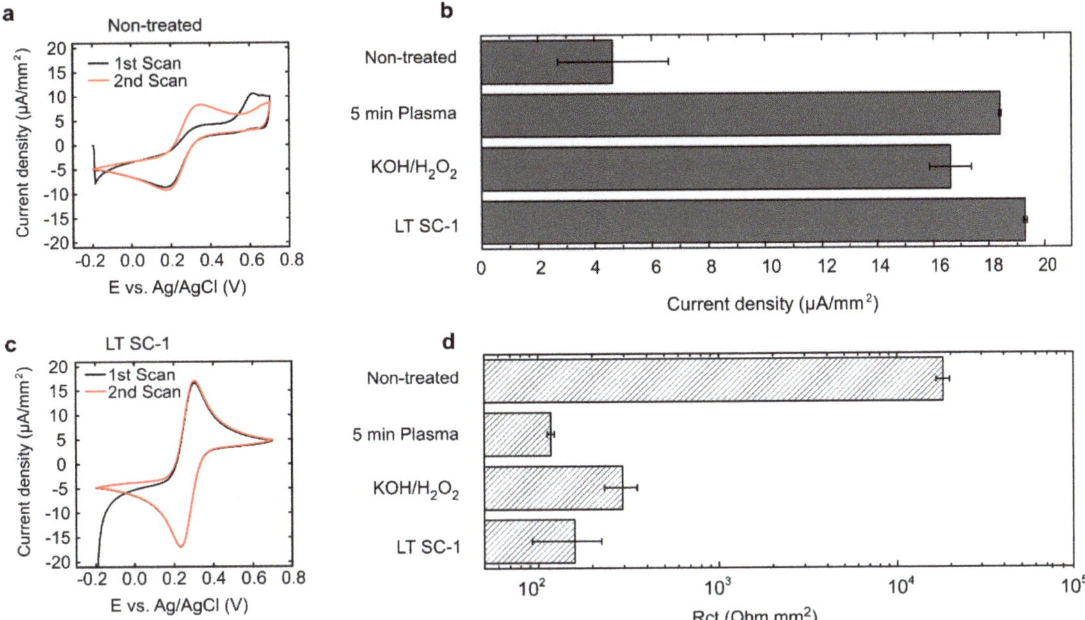

Figure 1. The effect of cleaning on electrode performance. (**a**) CVs in ferri-/ferrocyanide solution with non-treated PCB. (**b**) Comparison of current density obtained using CV in ferri-/ferrocyanide solution using multiple cleaning techniques by determining oxidation peak height. (**c**) CV in ferri-/ferrocyanide solution with PCB that underwent LT SC-1 cleaning treatment. (**d**) Comparison of Rct obtained using EIS in ferri-/ferrocyanide solution using multiple cleaning techniques. Bars represent the mean and error bars represent the SD, $N = 4$ in (**b**) and (**d**).

Previous reports revealed that commercially fabricated PCB electrodes are covered with an organic layer [5,9]. There are multiple possibilities of organic layer removal, such as oxygen plasma treatment [11], which is an interesting possibility due to the availability of the technique in PCB manufacturing plants and the low environmental impact [12]. Five-minute treatment in oxygen plasma and subsequent CV analysis revealed that the current density increased to 18.4 μA/mm^2 and the peak-to-peak separation decreased to 68 mV, indicating close-to-ideal behaviour for the ferri-/ferrocyanide couple. The average R_{ct} was 117 kΩmm^2 (Figure 1b,d), while peak-to-peak separation in non-treated PCBs was 234 mV, which indicated that the insulating layer was not removed fully; this was also confirmed by EIS, where R_{ct} was found to be over 18 kΩmm^2 (Figure 1d).

Another approach to removing organic contaminants on gold surfaces is a wet treatment with potassium hydroxide/hydrogen peroxide solution (50 mM KOH with 30% H_2O_2), as demonstrated previously [13]. CV and EIS analysis revealed 16.6 μA/mm^2 oxidation currents and R_{ct}, of 294 kΩmm^2 while peak-to-peak separation remained similar at 71 mV. This indicates that the KOH/H_2O_2 treatment is an appropriate but not ideal process for the removal of impurities.

Another strategy for the removal of organic contaminants used in the semiconductor industry is standard clean 1 (SC-1), the first part of a multi-step RCA cleaning procedure, developed at the Radio Corporation of America in 1965 [14]. SC-1 includes immersion of the wafer in a mixture of hydrogen peroxide and ammonium hydroxide at 80 °C. This mixture initiates oxidative breakdown and dissolution of metallic ions such as copper, nickel, and chromium. Ammonium hydroxide acts as a complexing agent, holding Cu ions in solution, while hydrogen peroxide is an oxidising agent dissolving metallic copper [15]. Unfortunately, the PCB silkscreen and solder mask can be affected by the highly active

SC-1 solution, so this process was adopted to clean PCBs by lowering the temperature of the SC-1 solution to RT and subsequent immersion of the PCB electrodes in acetone, isopropyl alcohol, and water to facilitate the complete removal of contaminants. This process will be referred to as low-temperature SC-1 clean (LT SC-1). Subsequent CV and EIS analysis in PCBs that underwent LT SC-1 clean revealed a consistent oxidation current density of 19.3 $\mu A/mm^2$, average peak-to-peak separation of 69 mV, and an average R_{ct}, of 158 Ωmm^2 (Figure 1b–d). These data indicate that all the abovementioned procedures can be applied for PCB surface cleaning. It should, however, be noted that the performance of electrochemical sensors will strongly be affected by electrode reproducibility; hence, further methods and method combinations were explored to ensure highly reproducible electrode characteristics (Figures S2–S4). Among these, potential cycling in sulphuric acid, which is a widely used approach for the electrochemical cleaning of gold surfaces [16], was also explored as a post-processing step and demonstrated after increased capability to remove the organic layer in PCB electrodes (Figure S5).

Closer examinations of the CV scans in H_2SO_4 revealed inconsistent behaviour of the PCB electrodes (Figure S6). CV scans showed consistent oxidation and reduction of gold but an unknown peak appearing in the region between 0 and 0.4 V vs. Ag/AgCl (KCl). This could be the oxidation of copper impurities [17,18] or exposure of Cu under the Ni and Au. During the electroplating process, a certain level of impurities remains in the gold plating solution [15]. Although the maximum allowed level of Cu impurities in the electroplating bath is relatively low, Cu impurities are present on the electrode surface and can remain there in small quantities even after the electrode cleaning [5]. To confirm the source of the impurities, a PCB electrode that did not exhibit impurity peaks was exposed to increasing concentrations of $CuSO_4$ in H_2SO_4 solution. Oxidation of Cu^{2+} ions was observed in the region of 0–0.3 V, presenting multiple peaks at high concentration (>300 μM); see Figure 2.

The observed peaks were integrated, and the peak area was plotted vs. the concentration of added $CuSO_4$, revealing a linear relationship and confirming that the observed peaks were due to Cu impurities (Figure 2a,b).

To evaluate the level of Cu impurities, CVs of cleaned PCB electrodes were performed in PBS at high scan rates of 1 V/s. This revealed very high Cu peaks in PCBs cleaned with plasma (not shown) and lower Cu peak in PCBs cleaned with KOH/H_2O_2 treatment or LT SC-1 clean. Furthermore, when KOH/H_2O_2-treated PCBs were also cleaned with electrochemical cycling in H_2SO_4 solution, the Cu impurity peak disappeared completely (Figure 2c). A more systematic examination of different cleaning combinations was then tested in an effort to determine the most optimal procedure. Plasma treatment, KOH/H_2O_2 cleaning, and LT SC-1 clean were examined in combination with CV cycling in H_2SO_4, CV cycling in KOH, and commercially available copper etchant solution.

After plasma exposure, the average Cu peaks were very high at 9.1 $\mu A/mm^2$ and dropped to 1.0 $\mu A/mm^2$ with H_2SO_4 cycling, 2.8 $\mu A/mm^2$ with KOH cycling, and 0.4 $\mu A/mm^2$ with exposure to Cu etchant (Figure 2d). The data suggest that Cu impurities are best removed with a wet process; hence, plasma alone would not be suitable for PCB cleaning but can be combined with a Cu etch step to achieve appropriate surface characteristics.

KOH/H_2O_2 treatment revealed average Cu peaks of 1.9 $\mu A/mm^2$, which dropped to below 1 $\mu A/mm^2$ with every subsequent step (Figure 2d). LT SC-1 cleaning revealed low and very consistent levels of Cu impurities, with average Cu peaks of 0.4 $\mu A/mm^2$. This confirms that a wet process is more suitable for the removal of Cu impurities.

To obtain greater insight into the reproducibility of the pre-treatment methodologies, assessment of a larger number of PCB electrodes was performed using two final candidates: LT SC-1 clean, due to its consistency, and KOH/H_2O_2 treatment followed by H_2SO_4 cycling, due to its excellent Cu removal capabilities. The latter procedure revealed almost complete removal of Cu peaks in some electrodes; however, other electrodes exhibited higher impurity peaks (Figure 3a). The range of impurities covered almost three orders of magnitude from 0.03 $\mu A/mm^2$ to 15 $\mu A/mm^2$ with an average of 1.6 $\mu A/mm^2$. On the other hand, LT SC-1 cleaning revealed a lower average impurity peak of 0.2 μA/mm^2 with the range from 0.1 μA/mm^2 to 0.8 μA/mm^2 using 120 PCB electrodes.

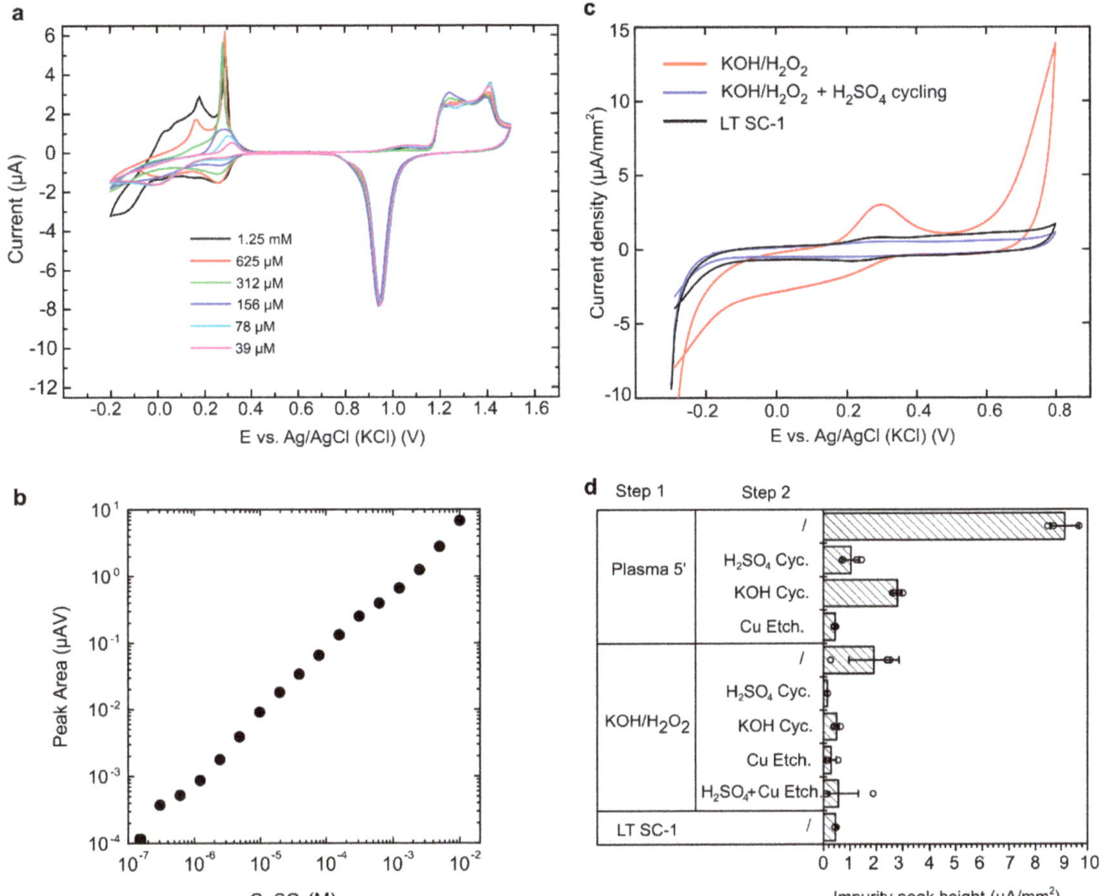

Figure 2. Determination of copper impurities in PCB electrodes. (**a**) CV curves obtained in sulphuric acid with increasing concentrations of $CuSO_4$. (**b**) The relationship between $CuSO_4$ concentration and peak area obtained in CV scans. (**c**) Analyses by CV in PBS, where Cu impurity peak is revealed at approximately 0.3 V. (**d**) The height of impurity peaks in PCBs that underwent various combinations of cleaning procedures (step 1 plus step 2). Bars represent the mean, error bars represent the SD, and empty circles represent individual data points.

Additionally, Cu peak behaviour during H_2SO_4 cycling was not consistent. In some electrodes, Cu peaks decreased during CV cycling, while an increase in Cu peaks was observed with other PCBs. This indicated that two phenomena were occurring simultaneously.

We propose that this could be explained by copper contamination of the electroplated gold layer, depicted in Figure 3b–d.

Commercially fabricated PCBs are delivered with a layer of organic impurities, which prevent direct oxidation of ferri-/ferrocyanide couple or, more importantly, the formation of self-assembly monolayers (SAMs) and conjugation of capturing probes on the PCB surface (Figure 3b). Additionally, Cu is present in the initial layer [5], which can cause electroactivity of the PCB electrodes and can mask other desired processes occurring on the electrodes. A wet KOH/H_2O_2 treatment removed organic and the bulk of the inorganic impurities, with some remaining on the surface (Figure 3c), leading to Cu peaks observed in CV in PBS. Further CV cycling in sulphuric acid promotes the removal of Cu impurities, which can be seen as a cycle-dependent decrease in Cu peaks over time. However, during the scan, a thin layer of gold can be dissolved in the solution [19,20]. We postulate that this can expose Cu impurities in the interior of the gold layer, which is there as a consequence of the partly contaminated gold plating solution during the gold electrodeposition and was previously 'hidden' from the surface, protected from the wet cleaning process (Figure 3d). This is an increase in Cu peaks over time, observed by continuous scanning of PCBs in sulphuric acid. Besides the chemical surface composition, electrode roughness can possess a great effect on the electrochemical sensor's performance [21–23].

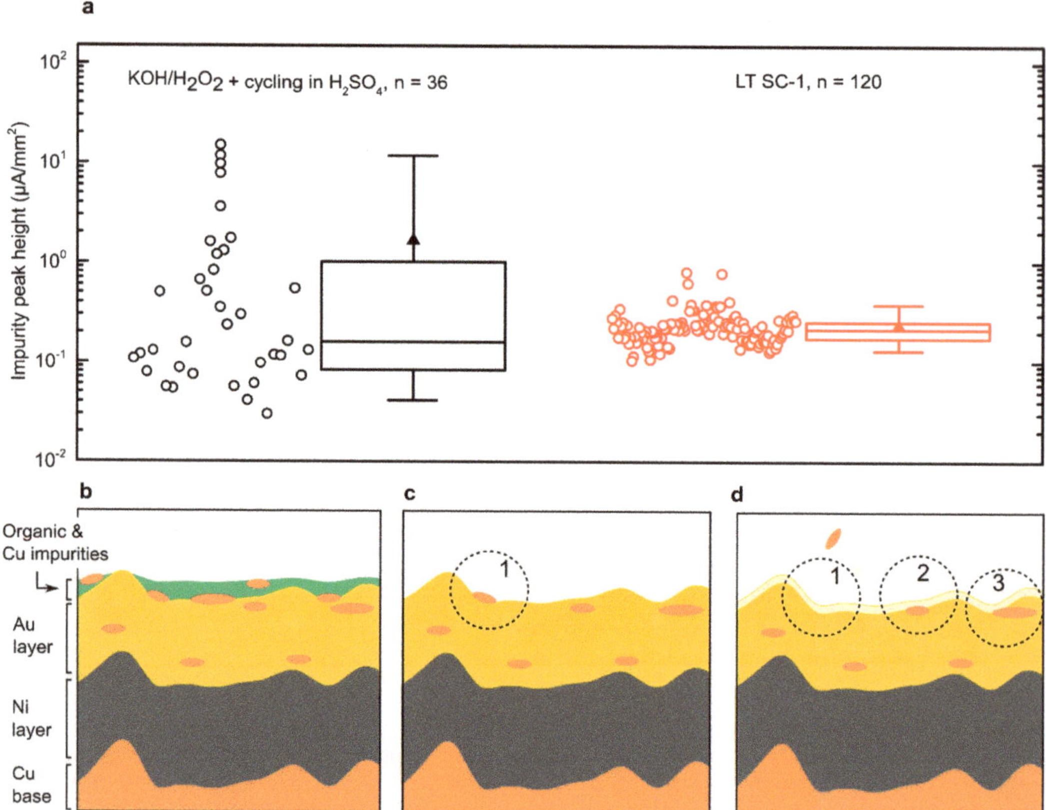

Figure 3. Electroactive impurity peaks in two different cleaning procedures and conceptual representation of PCB impurity removal process. (**a**) The height of impurity peaks in the larger sample number for two promising cleaning procedures. Empty circles represent individual electrodes, lines in a box represent the median, 25th and 75th percentile, the triangle is the mean value, whiskers represent the 5th and 95th percentile. (**b**) Non-treated PCB consisting of layers of a copper base, electroplated nickel, and gold. Cu impurities are found on top of the gold and within the gold layer. (**c**) PCB surface after wet KOH/H_2O_2 cleaning with removed organic and inorganic impurities. (**d**) Removal of the thin gold layer during CV cycling. Circle 1 represents the removal of Cu impurity and circles 2 and 3 represent newly formed exposure of the hidden Cu impurities within the plated electrodes.

3.2. Evaluation of PCB Surface Roughness

Extensive research can be found on strategies to increase the gold surface roughness to achieve high surface-to-volume ratios and enhance charge transfer properties [24].

However, high surface roughness can result in increased fouling properties [22,25,26]. Planar electrodes such as evaporated gold on glass or silicon substrates benefit from the smoothness of their respective substrates, leading to controllable smooth surfaces [27]. Although some applications utilise planar gold electrodes, a low-cost alternative solution predominantly includes screen-printed electrodes, which are known to possess very high surface roughness, due to the nature of their production [21,23]. PCB processing is expected to produce a surface of intermediate roughness, compared to these two commonly used technologies.

To evaluate PCB electrodes, surface roughness was studied using atomic force microscopy as well as evaluated electrochemically. A thin-film, thermally evaporated planar gold electrode was compared to PCB electrodes. AFM revealed a smooth surface in planar electrodes, where all features fell within the 11 nm range. The average root mean square roughness from three measurements was 763.4 pm ± 62.8 pm. Five PCB electrodes were analysed with AFM, revealing the average root mean square roughness to be 20.99 nm ± 3.98 nm. This demonstrates that PCB electrodes are over 25-times rougher than planar gold electrodes. Representation of individual AFM scans can be seen in Figure 4a,b. Comparison of the extracted profiles revealed that PCB electrodes include features up to 100 nm in height (Figure 4c), compared to few-nm-level features in a planar gold electrode.

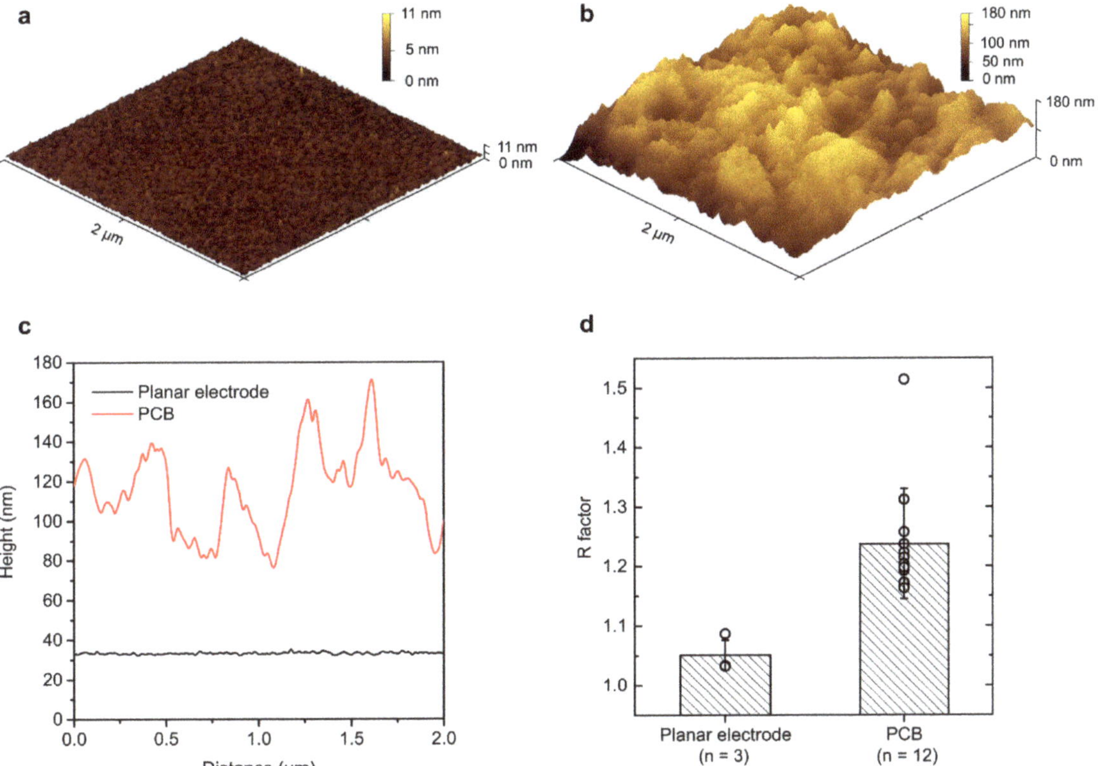

Figure 4. PCB surface roughness evaluation. (**a**) Example of the AFM profile for planar gold electrodes and (**b**) PCB electrodes. (**c**) The obtained profile from both representative samples. (**d**) Roughness factors obtained by CV scanning in H_2SO_4. Bars represent the mean, error bars represent the SD, and empty circles represent individual electrode datapoints.

Surface characteristics were then also evaluated electrochemically. By CV cycling in sulphuric acid solution, the electrode's electroactive surface area was evaluated and the roughness factor was calculated. The planar gold electrode revealed a mean roughness factor of 1.05, while the PCB electrode had an average roughness factor of 1.24 (Figure 4d). This is a small difference considering that the AFM results show 25-times higher surface roughness. Two factors contribute to the level of gold electroactivity: surface roughness and gold availability. Increased surface roughness increases the available area, but impurities covering the gold prevent direct contact with the solution and prevent gold oxidation, therefore decreasing the available surface area. Hence, electrochemical evaluations of PCB electrodes should be taken with caution, as small roughness factors do not necessarily indicate smooth surfaces but could be a consequence of impurities and decreased gold availability.

3.3. Electrochemical Protein Quantification Using Commercial PCB Electrodes

To demonstrate that commercial PCB electrodes can be used for the sensitive quantification of protein biomarkers, procalcitonin, a promising sepsis biomarker [28,29], was used as a model assay. PCT-based ELISA was constructed on a 96-well plate and TMB was measured optically and electrochemically (Figure 5).

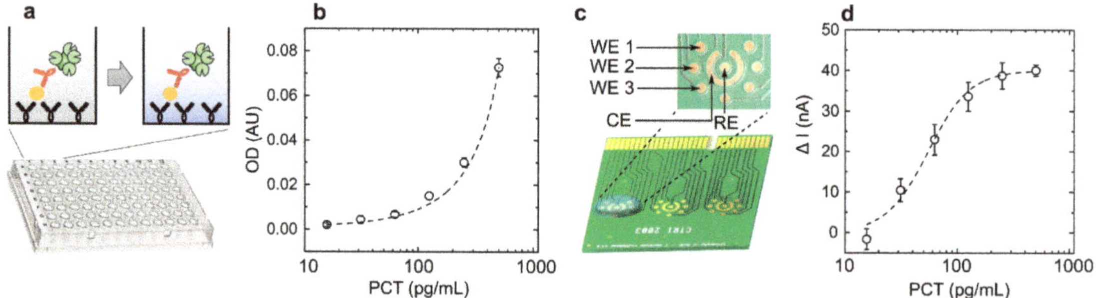

Figure 5. Optical and electrochemical quantification of PCT ELISA assay. (**a**) Conceptual representation of the ELISA assay performed in a 96-well plate, where the addition of TMB is seen as a colour change. (**b**) Optical detection of the colour change ($N = 3$). (**c**) Electrochemical set-up with gold PCB working electrodes (WE 1, 2, and 3) and shared counter (CE) and reference (RE) electrode. (**d**) Electrochemical detection of TMB using chronoamperometry ($N = 3$).

Optical detection revealed an exponential increase in OD with increasing PCT concentrations, while electrochemical quantification revealed large changes in currents at low PCT concentrations, with an LoD of 31 pg/mL (or 2 pM). This demonstrated that the commercially manufactured PCB electrodes can be used for the sensitive electrochemical detection of catalysed TMB and consequently any protein when analysed with ELISA-based assays. This could be further expanded and integrated into microfluidic devices for low-cost quantification of any biomarker of interest, which can be quantified using ELISA-based systems.

Although the above system can be used for the quantification of an ELISA-based assay, many biosensors require immobilisation of the capturing probe on the electrode where the measurement is performed. An example of such a sensor is a label-free DNA sensor, presented in the next section.

3.4. Detection of DNA Using Commercial PCB Electrodes

To evaluate the feasibility of capture probe immobilisation on commercially fabricated PCB electrodes, a biosensor was fabricated using single-stranded peptide nucleic acids (ssPNA) as biological probes for the detection of genomic *E. coli* ssDNA. The sequences came from genes that code for an *E. coli*-specific virulent factor called fimbriae protein, which aids in the invasion of the host system and is highly conserved within the O157:H7 serotype genome [30]. Thiolated PNA probes were immobilised on pre-cleaned commercial PCB electrodes and the electrochemical impedance spectra were recorded upon measurement of a blank as well as five concentrations of ssDNA (Figure 6). The electrochemical impedance spectra data show a significant change between 0.01 and 100 nM. The non-linear curve fit shows a significant R-squared value of 0.95. The limit of detection (LoD) was calculated to be 110 pM using an optimised co-immobilisation molar ratio of 1:15. The Nyquist plot shows a dynamic range between 100 pM and 10 nM, with an observed saturation effect between 10 and 100 nM.

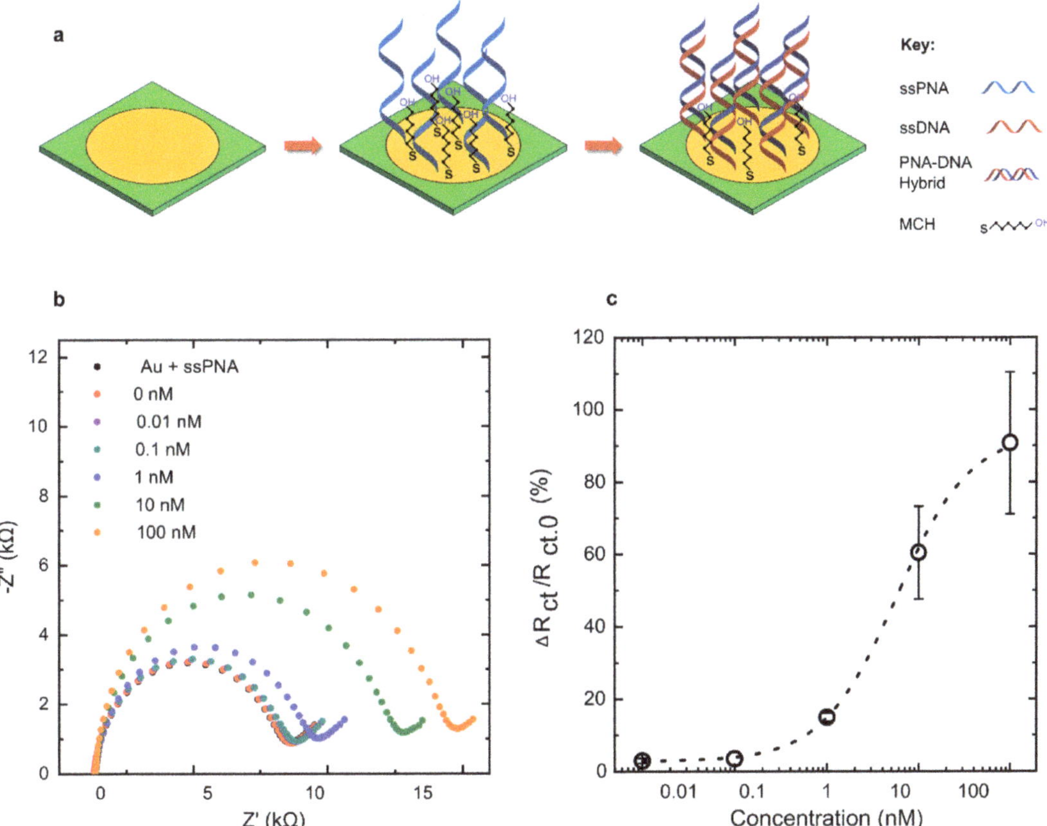

Figure 6. EIS-based PNA–DNA hybridisation assay on commercial PCB electrodes. (**a**) PCB gold electrode functionalised first with thiolated PNA probes and co-immobilised with MCH spacer molecules before direct hybridisation with ssDNA samples. (**b**) Typical Nyquist plots obtained for PNA–DNA hybridisation using thiol surface chemistry on PCB with five concentrations of target ssDNA and blank sample. (**c**) A calibration curve for the percentage change in R_{ct} using five concentrations of ssDNA in buffer ($N = 3$).

We believe that the higher error at high concentrations of the DNA target can be attributed to variation in the surface roughness and immobilised probe density. This seems to be less prominent at a lower concentration of target DNA, as a highly reproducible change in Rct (approximately 20%) can be seen at 1 nM DNA concentration. Due to the surface morphology leading to variability in SAM formation, the exact probe density varied from electrode to electrode. As higher concentrations of target DNA were introduced to the sample, the sensor surface began to saturate, and the probe density can have a major effect on the obtained Rct. This can be seen as a larger variation in the sensor response at a high target concentration.

4. Conclusions

The use of commercially employed processes for PCB manufacturing in constructing sensitive and reliable electrochemical sensors could enable low-cost diagnostic microsystems, which could be scaled and produced locally, using existing manufacturing infrastructure. Nevertheless, this required reliability of electrochemical biosensors is underpinned by the reproducible physicochemical characteristics of the sensing electrodes and the respective processes used for their construction. Gold bioavailability and the lack of contaminants in electrode surfaces are crucial in the final device performance as they directly affect the surface chemistry and the final readout of the sensor. Detailed electrochemical characterisation revealed suitable electrode cleaning procedures, which provide reproducible electrode characteristics and effective removal of electroactive contaminants.

Gold-plated PCB electrodes have a considerably higher surface roughness (RMS roughness of ca. 20 nm) compared to cleanroom-fabricated thin-film gold electrodes (RMS roughness of ca. 1 nm) but, at the same time, orders of magnitude smaller than the widely used screen-printed ones (RMS roughness of 1 μm) [24]. As surface roughness is known to have a profound effect on biosensor performance, this indicates that sensors that cannot be implemented in commercially screen-printed electrodes due to their high surface roughness can be implemented on PCB electrodes, offering a low-cost alternative to the high-specification evaporated ones.

Finally, reliable electrochemical detection using such optimised, commercial PCB sensing electrodes was demonstrated. Label-free detection of DNA based on PNA probes is shown, along with quantification of the protein biomarker PCT, achieving clinically relevant limits of detection. With this knowledge, electrochemical sensors can now be constructed on commercially manufactured PCB electrodes for a multitude of biomarker quantification applications, paving the way for reliable commercial Lab-on-PCB biomedical diagnostic platforms.

Supplementary Materials: The following are available online at https://www.mdpi.com/article/10.3390/mi12070793/s1, Table S1, Additional cleaning steps used in combination with previously described cleaning procedures. Figure S1, PCB boards used to characterise PCB electrodes. The PCB with eight electrodes (WEs) which connect to a PCI express slot-type connector. Figure S2, Current obtained from CV in ferri-/ferrocyanide with a combination of multiple cleaning steps. Bars represent the average and error bars represent the SD, $N = 4$. Figure S3, Peak-to-peak separation value obtained from CV in ferri-/ferrocyanide with a combination of multiple cleaning steps. Bars represent the average and error bars represent the SD, $N = 4$. Figure S4, Charge transfer resistance values obtained from EIS in ferri-/ferrocyanide solution with a combination of multiple cleaning steps. Bars represent the average and error bars represent the SD, $N = 4$. Figure S5, Charge transfer resistance obtained with PCB electrodes in ferri-/ferrocyanide solution with different treatments. Bars represent the mean and error bars represent the SD, $N = 4$. Figure S6, Examples of CV curves obtained by electrochemical polishing of PCB electrodes in sulphuric acid. Ox represents the peak demonstrating formation of gold oxide layer and Red represents the reduction of gold oxide in a reverse CV scan.

Author Contributions: All authors contributed to the writing of the manuscript. U.Z. designed the study and performed the experiments under the direction of P.E. and D.M.; J.R. designed and performed the experiments employing PNA-based DNA sensors under the direction of P.E. All authors have read and agreed to the published version of the manuscript.

Funding: This work was supported by the University of Bath and the Rosetrees Trust (project M681) as well as the UK Natural Environment Research Council (NERC) GW4 FRESH CDT.

Data Availability Statement: Data available from authors upon request.

Conflicts of Interest: The authors declare no competing financial interests.

References

1. Güth, F.; Arki, P.; Löher, T.; Ostmann, A.; Joseph, Y. Electrochemical Sensors Based on Printed Circuit Board Technologies. *Procedia Eng.* **2016**, *168*, 452–455. [CrossRef]
2. Shamkhalichenar, H.; Bueche, C.; Choi, J.-W. Printed Circuit Board (PCB) Technology for Electrochemical Sensors and Sensing Platforms. *Biosensors* **2020**, *10*, 159. [CrossRef] [PubMed]
3. Lammerink, T.; Spiering, V.; Elwenspoek, M.; Fluitman, J.; Berg, A.V.D. Modular concept for fluid handling systems. A demonstrator micro analysis system. In Proceedings of the Ninth International Workshop on Micro Electromechanical Systems, San Diego, CA, USA, 11–15 February 1996; Institute of Electrical and Electronics Engineers (IEEE): Piscataway, NJ, USA, 1996.
4. Moschou, D.; Tserepi, A. The lab-on-PCB approach: Tackling the μTAS commercial upscaling bottleneck. *Lab Chip* **2017**, *17*, 1388–1405. [CrossRef] [PubMed]
5. Jolly, P.; Rainbow, J.; Regoutz, A.; Estrela, P.; Moschou, D. A PNA-based Lab-on-PCB diagnostic platform for rapid and high sensitivity DNA quantification. *Biosens. Bioelectron.* **2019**, *123*, 244–250. [CrossRef] [PubMed]
6. Land, K.J.; Boeras, D.I.; Chen, X.-S.; Ramsay, A.R.; Peeling, R.W. REASSURED diagnostics to inform disease control strategies, strengthen health systems and improve patient outcomes. *Nat. Microbiol.* **2019**, *4*, 46–54. [CrossRef]
7. Yetisen, A.K.; Ali, K.; Akram, M.S.; Lowe, C.R. Paper-based microfluidic point-of-care diagnostic devices. *Lab Chip* **2013**, *13*, 2210–2251. [CrossRef] [PubMed]
8. Burke, L.D.; Buckley, D.T.; Morrissey, J.A. Novel view of the electrochemistry of gold. *Analyst* **1994**, *119*, 841–845. [CrossRef]
9. Dutta, G.; Jallow, A.A.; Paul, D.; Moschou, D. Label-Free Electrochemical Detection of S. mutans Exploiting Commercially Fabricated Printed Circuit Board Sensing Electrodes. *Micromachines* **2019**, *10*, 575. [CrossRef]
10. Moschou, D. Amperometric IFN-gamma immunosensors with commercially fabricated PCB sensing elec-trodes. *Biosens. Bioelectron.* **2016**, *86*, 805–810. [CrossRef]
11. Raiber, K.; Terfort, A.; Benndorf, C.; Krings, N.; Strehblow, H.-H. Removal of self-assembled monolayers of alkanethiolates on gold by plasma cleaning. *Surf. Sci.* **2005**, *595*, 56–63. [CrossRef]
12. Fierro, L.; Getty, J. Plasma Processes for Printed Circuit Board Manufacturing. 1 January 2003. Available online: https://www.yumpu.com/en/document/read/626798/plasma-processes-for-printed-circuit-board-manufacturing (accessed on 10 April 2020).
13. Heiskanen, A.; Spégel, C.F.; Kostesha, N.; Ruzgas, T.; Emnéus, J. Monitoring of Saccharomyces cerevisiae Cell Proliferation on Thiol-Modified Planar Gold Microelectrodes Using Impedance Spectroscopy. *Langmuir* **2008**, *24*, 9066–9073. [CrossRef] [PubMed]
14. Kern, W. The Evolution of Silicon Wafer Cleaning Technology. *J. Electrochem. Soc.* **1990**, *137*, 1887–1892. [CrossRef]
15. Goosey, M. Printed Circuits Handbook. *Circuit World* **2010**, *36*. [CrossRef]
16. Tkac, J.; Davis, J.J. An optimised electrode pre-treatment for SAM formation on polycrystalline gold. *J. Electroanal. Chem.* **2008**, *621*, 117–120. [CrossRef]
17. Labuda, A.; Paul, W.; Pietrobon, B.; Lennox, R.B.; Grutter, P.H.; Bennewitz, R. High-resolution friction force microscopy under electrochemical control. *Rev. Sci. Instrum.* **2010**, *81*, 83701. [CrossRef]
18. Wahl, A.; Dawson, K.; Sassiat, N.; Quinn, A.J.; O'Riordan, A. Nanomolar Trace Metal Analysis of Copper at Gold Microband Arrays. *J. Phys. Conf. Ser.* **2011**, *307*, 012061. [CrossRef]
19. Cherevko, S.; Topalov, A.A.; Zeradjanin, A.R.; Katsounaros, I.; Mayrhofer, K. Gold dissolution: Towards understanding of noble metal corrosion. *RSC Adv.* **2013**, *3*, 16516–16527. [CrossRef]
20. Ho, L.S.J.; Limson, J.L.; Fogel, R. Certain Methods of Electrode Pretreatment Create Misleading Responses in Impedimetric Aptamer Biosensors. *ACS Omega* **2019**, *4*, 5839–5847. [CrossRef]
21. Butterworth, A.; Blues, E.; Williamson, P.; Cardona, M.; Gray, L.; Corrigan, D.K. SAM Composition and Electrode Roughness Affect Performance of a DNA Biosensor for Antibiotic Resistance. *Biosensors* **2019**, *9*, 22. [CrossRef]
22. Hoogvliet, J. Electrochemical pretreatment of polycrystalline gold electrodes to produce a reproducible sur-face roughness for self-assembly: A study in phosphate buffer pH 7.4. *Anal. Chem.* **2000**, *72*, 2016–2021. [CrossRef] [PubMed]
23. Obaje, E.A. Carbon screen-printed electrodes on ceramic substrates for label-free molecular detection of anti-biotic resistance. *J. Interdiscip. Nanomed.* **2016**, *1*, 93–109. [CrossRef]
24. Collinson, M. Nanoporous Gold Electrodes and Their Applications in Analytical Chemistry. *ISRN Anal. Chem.* **2013**, *2013*, 1–21. [CrossRef]
25. Benites, T.A. Efeitos da rugosidade superficial nas propriedades de passivação de monocamadas orgânicas automontadas. *Quim. Nova* **2014**, *37*, 1533–1537.
26. Santos, A.; Piccoli, J.P.; Santos-Filho, N.A.; Cilli, E.M.; Bueno, P.R. Redox-tagged peptide for capacitive diagnostic assays. *Biosens. Bioelectron.* **2015**, *68*, 281–287. [CrossRef] [PubMed]
27. Mahmoodi, N.; Rushdi, A.I.; Bowen, J.; Sabouri, A.; Anthony, C.J.; Mendes, P.M.; Preece, J.A. Room temperature thermally evaporated thin Au film on Si suitable for application of thiol self-assembled monolayers in micro/nano-electro-mechanical-systems sensors. *J. Vac. Sci. Technol. A* **2017**, *35*, 041514. [CrossRef]
28. Chan, T.; Gu, F. Early diagnosis of sepsis using serum biomarkers. *Expert Rev. Mol. Diagn.* **2011**, *11*, 487–496. [CrossRef] [PubMed]
29. Faix, J.D. Biomarkers of sepsis. *Crit. Rev. Clin. Lab. Sci.* **2013**, *50*, 23–36. [CrossRef] [PubMed]
30. Saeedi, P.; Yazdanparast, M.; Behzadi, E.; Salmanian, A.H.; Mousavi, S.L.; Nazarian, S.; Amani, J. A review on strategies for decreasing *E. coli* O157:H7 risk in animals. *Microb. Pathog.* **2017**, *103*, 186–195. [CrossRef]

 *micromachines*

Article

Biocompatibility Study of a Commercial Printed Circuit Board for Biomedical Applications: Lab-on-PCB for Organotypic Retina Cultures

Jesús David Urbano-Gámez [1], Lourdes Valdés-Sánchez [2], Carmen Aracil [1,*], Berta de la Cerda [2,*], Francisco Perdigones [1], Álvaro Plaza Reyes [2], Francisco J. Díaz-Corrales [2], Isabel Relimpio López [3] and José Manuel Quero [1]

1 Electronic Technology Group, Department of Electronic Engineering, Higher Technical School of Engineering, University of Seville, Avda. de los Descubrimientos sn, 41092 Seville, Spain; jurbano1@us.es (J.D.U.-G.); fperdigones@us.es (F.P.); quero@us.es (J.M.Q.)
2 Department of Regeneration and Cell Therapy, Andalusian Molecular Biology and Regenerative Medicine Centre (CABIMER), Avda. Américo Vespucio 24, 41092 Seville, Spain; lourdes.valdes@cabimer.es (L.V.-S.); alvaro.plaza@cabimer.es (Á.P.R.); francisco.diaz@cabimer.es (F.J.D.-C.)
3 RETICS Oftared, Carlos III Institute of Health (Spain), Ministry of Health RD16/0008/0010, University Hospital Virgen Macarena, Avda. Dr. Fedriani, 3, 41009 Seville, Spain; irelimpio@gmail.com
* Correspondence: caracil@gte.esi.us.es (C.A.); berta.delacerda@cabimer.es (B.d.l.C.)

Abstract: Printed circuit board (PCB) technology is well known, reliable, and low-cost, and its application to biomedicine, which implies the integration of microfluidics and electronics, has led to Lab-on-PCB. However, the biocompatibility of the involved materials has to be examined if they are in contact with biological elements. In this paper, the solder mask (PSR-2000 CD02G/CA-25 CD01, Taiyo Ink (Suzhou) Co., Ltd., Suzhou, China) of a commercial PCB has been studied for retinal cultures. For this purpose, retinal explants have been cultured over this substrate, both on open and closed systems, with successful results. Cell viability data shows that the solder mask has no cytotoxic effect on the culture allowing the application of PCB as the substrate of customized microelectrode arrays (MEAs). Finally, a comparative study of the biocompatibility of the 3D printer Uniz zSG amber resin has also been carried out.

Keywords: Lab-on-PCB; printed circuit board; organotypic culture; biomedical applications

Citation: Urbano-Gámez, J.D.; Valdés-Sánchez, L.; Aracil, C.; de la Cerda, B.; Perdigones, F.; Plaza Reyes, Á.; Díaz-Corrales, F.J.; Relimpio López, I.; Quero, J.M. Biocompatibility Study of a Commercial Printed Circuit Board for Biomedical Applications: Lab-on-PCB for Organotypic Retina Cultures. *Micromachines* **2021**, *12*, 1469. https://doi.org/10.3390/mi12121469

Academic Editors: Nam-Trung Nguyen and Angeliki Tserepi

Received: 13 October 2021
Accepted: 26 November 2021
Published: 29 November 2021

1. Introduction

Although printed circuit board (PCB) substrates are typically used for electronics, the development of devices based on PCB substrates has been the subject of increasing research over the years [1,2]. The reason for this lies in the advantages that PCB implies for biomedical applications and marketability [3,4], such as low cost, commercial availability, and the possibility of integrating electronic circuits with ease. It is important to note that this kind of device combined with microfluidics can perform different laboratory tasks, such as reactions or mixing in a small laboratory with the dimensions of a credit card [4]. Moreover, the possibility to integrate heaters and temperature sensors can provide a chamber with the optimal biological conditions. These devices are named PCB-based lab-on-a-chip or Lab-on-PCB (LoP) devices. Up to now, these devices have been applied to several biomedical fields, such as detection of cell viability [5], molecular diagnosis [6], DNA amplification [7,8], and electrolytes detection [9].

Focusing on one of the most important biomedical applications, electrophysiology, where MEAs are frequently used, different approaches have been developed regarding the design of these devices. Electrostimulation and acquisition of signals from the biological material is accomplished thanks to these MEAs, which have to be just below the cells or organotypic cultures. The insulation material is an important component of the MEAs, since it is in direct contact with the culture. The role of this material is to avoid the electrical contact between cells/tissues and the metal tracks, releasing only the electrodes. A typical insulation material is silicon nitride (SiN), used in many MEAs, for instance, in the standard MEA of multichannel systems [10,11], in the high-density version [12], in the 3D version [13], and in the PEDOT-CNT MEA [14]. In the perforated versions of those MEAs, the insulation material is polyimide [15].

The PCB technology can be a good choice for performing these devices, in both rigid or flexible substrates, although the biocompatibility has to be considered. There are several examples of MEAS manufactured with this technology: a flexible one, using polyimide as both substrate and insulation material [16], and a fabricated one, using PCB (Eco MEA) with Elpemer®2467 or PSR-4000 GP01EU (Peters, Kempen, Germany and Taiyo America, Carson City, NV, USA, respectively) [17] as insulation material. Other designs can be found in state of the art such as the PCB-based 3D MEA described on [18]. This MEA has gold electrodes covered with SU-8 as an insulation layer. A different version of PCB-based MEA is reported on [19], where 3D gold microelectrodes are embedded on polydimethylsiloxane (PDMS). The biocompatibility of this MEA was demonstrated for retinal explant from a retinitis pigmentosa mouse-model. However, in the latter examples the fabrication process is quite complex.

Our goal is the fabrication of a MEA on PCB for specific biomedical applications, which is the evaluation of neuroprotective strategies for retinal degenerative diseases, and we expect to achieve an easy, reliable, and low-cost design that can be adjusted to a wide range of applications. The objective is to give the researchers the possibility to develop their customized MEAs, at low cost, instead of being limited to the current commercial MEAs or investing a higher cost for suggesting a design. Regarding the materials, the rigid PCB substrate is composed of copper and FR4, where the copper is typically covered with gold. Standard and commercial PCB technology includes a solder mask as an insulation layer. The specific insulation layer depends on the PCB manufacturing company. Polyimide is a biocompatible material used as a flexible PCB insulation layer, particularly for retinal implants [20]. In addition, a typical solder mask of flexible substrates (PSR-9000 FLX03G LDI, Taiyo America, Carson City, NV, USA) seems to be biocompatible [21]. However, we plan to use commercial rigid PCB substrates for manufacturing MEAs, from a specific company [22]. This company uses specifically a white solder mask (PSR-2000 CD02G/CA-25 CD01, Taiyo Ink (Suzhou) Co., Ltd., Suzhou, China) [22], which has to be tested for biocompatibility, which is the main aim of this work.

Regarding biocompatibility, it is convenient to perform the study for the specific application that will be implemented. In our case, the MEA will be applied to the study of retinal degenerative diseases. Retinal degenerative diseases range from classic genetic diseases such as retinitis pigmentosa to complex traits that constitute most of the blinding diseases in the industrialized world such as macular degeneration, glaucoma, and diabetic retinopathy. In every case, neurodegeneration and cell death ultimately produce low vision or blindness. The high impact of sight loss at the individual and societal level generates a demand for knowledge on the disease pathways and the development of new therapeutic approaches. Researchers look for neuroprotection to fight cell death and regenerative medicine methods to improve vision.

Organotypic culture has been applied to the study of retinal biology, including postnatal retinal development, factors influencing the retinal degenerative process, and the assay of candidate therapeutic molecules that may modify the degenerative process. The current retinal organotypic culture methodology was described for neonatal [23] mice retinas and adapted to adult mice retinas by Müller et al. [24]. The utility of the organotypic methodology benefits from the similarity of the retinal explant to the natural tissue compared to the cell culture of isolated retinal components while allowing the researchers to keep a tightly controlled experimental condition that sometimes is not possible in in vivo animal experiments. Additionally, an organotypic culture allows for continuous monitoring of the tissue via integrated optics and/or MEAs. Previously, some studies achieved electrical stimulation and recording applied to therapeutic evaluation, viability assessment, and functional characterization of retinal explants and organoids [19,25–29] with the use of MEAs as a fundamental tool.

It is also important, for the adequate design of the biocompatibility test, to consider the microfluidic platform where the organotypic culture will be located. In our case, the MEA will be integrated inside a lab-on-chip (LoC), so the biocompatibility tests have to be designed taking into account the organotypic culture that will be located inside a closed system. The standard retinal organotypic culture is based on a permeable Millicell membrane as support of the tissue, in which the retinas stay in a liquid–air interface. In our study, we have first compared the performance of two different open/static culture systems, with the Millicell membrane as a control, to evaluate the biocompatibility of the materials. In a second approach, we have tested the viability of the retinal tissue in the same two systems but in closed configuration, which is more similar to our goal of a stand-alone LoC system.

In this paper, a Lab-on-PCB for long-term organotypic retinal explant culture has been developed and used to demonstrate the biocompatibility of commercial rigid PCB substrates, specifically for white solder masks of the company JLCPCB (PSR-2000 CD02G/CA-25 CD01, Taiyo Ink (Suzhou) Co., Ltd., Suzhou, China) [22]. The ability to sustain the biological material in culture onto the commercial PCB implies the possibility of customization of MEAs with different dimensions, the number of electrodes and different shapes, and the use of an open-source and user-friendly software, just to be ordered to a fabrication company.

2. Culture Chambers

In order to study the biocompatibility of the solder mask, a PCB culture chamber, completely covered with solder mask, and with a frame of PMMA, has been manufactured. For comparative purposes, a different chamber has been also designed. In this case, it is based on a commercial glass microelectrode array (MEA), plus a frame of 3D printer resin. These culture platforms are described below.

2.1. PMMA/PCB Culture Chamber

This platform was designed using Klayout open-source software. The chamber dimensions, adapted to the MEA, correspond to 32 × 32 mm.

The culture chamber was composed of a cover and a base, both of them made of PMMA over a PCB substrate. This substrate was completely covered with an insulation layer, a solder mask, that was be in direct contact with the biological material, so it was mandatory to study its biocompatibility. The PCB substrate was a sheet with a height of 1.6 mm, which was covered, entirely, with a white ink solder mask (PSR-2000 CD02G/CA-25 CD01, Taiyo Ink (Suzhou) Co., Ltd., China) [22] used as an insulation layer. These parts can be seen in Figure 1A.

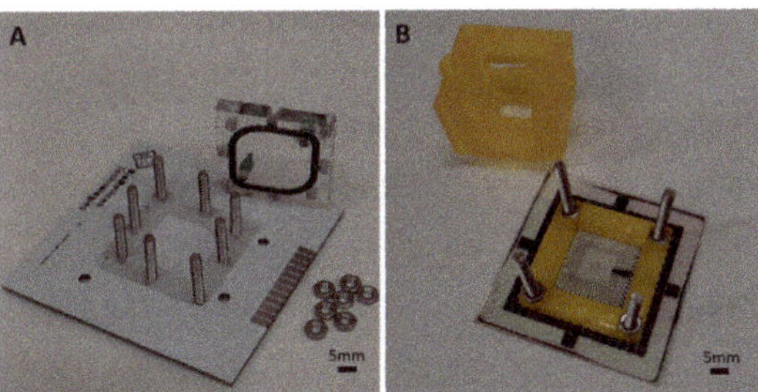

Figure 1. Culture chambers. (**A**) Culture chamber composed of PCB as substrate and a fluidic system made of PMMA. (**B**) Culture chamber composed of MEA from MultiChannel systems (MCS) as substrate and a fluidic system made of Uniz zSG amber resin.

The base of this platform was designed as an eye-shaped inner hole with a diameter of 20 mm and eight additional holes with 3 mm diameter, distributed around the external frame, for screws. These screws were used to ensure the correct alignment and coupling of both the cover and the base.

The cover includes two holes with a diameter of 2.7 mm to be used as inlet and outlet ports. Male mini-luer fluid connectors were inserted in these holes of the cover. In this part of the platform, an eye-shaped milled channel was also included where a sealing rubber ring (1.6 mm diameter) was attached to ensure correct isolation of the inner chamber and to avoid any leakage of liquid. In addition, eight holes of 3 mm diameter were added at the same position as the ones in the base, in order to align both parts of the platform with screws.

Regarding the fabrication process, an automated CO_2 laser milling machine was used for fabricating both the cover and the base, and 2 mm-thick and 5 mm-thick PMMA sheets were used to build the cover and the base, respectively. Once the screws were coupled to the base, the PCB substrate was attached using a biocompatible epoxy glue, EPO-TEK®301 (Epoxy Technology, Billerica, MA, USA). This adhesive was also used for the attachment of the sealing ring to the eye-shape milled channel of the cover.

2.2. *3D Printer Resin/Commercial Glass MEA Culture Chamber*

The design and manufacturing of this fluidic platform using a 3D printer resin have been previously published [30]. Briefly, a 3D model for organotypic cultures composed of two parts was designed using Fusion 360 software from Autodesk. The base and the fluidic platform were coupled with screws, for its integration to a commercial MEA. These parts can be seen in Figure 1B. The fluidic platform has two integrated connectors: inlet and outlet ports. These ports communicate with the culture chamber at different heights, that is, the output port is placed higher than the inlet port. This configuration creates a tilted top surface with an angle of 10° between both openings that helps the removal of air bubbles. The culture area was defined to be 21 × 21 mm, and the external dimensions of the base were 32 × 32 mm.

Regarding the manufacturing process, the Anycubic Photon 3D printer was used with the Uniz zSG Amber resin. This material will be in direct contact with the culture media, where the retinas are located, so its biocompatibility has also to be studied, but it will not be in direct contact with the retinas, so the biocompatibility test should replicate this specific situation. The isolation of the culture chamber was ensured by attaching a glass coverslip (tilted surface) in the window on the fluidic structure and using an ethylene-vinyl acetate (EVA) film, between the base and the fluidic platform. The adhesion of the base to the MEA surface and the attachment of the glass coverslip to the fluidic platform were performed using EPO-TEK®301 (Epoxy Technology, Billerica, MA, USA).

The substrate used for this culture platform was the commercial MEA 60MEA500/30iR-Ti (MultiChannel Systems MCS GmbH, Reutlingen, Germany). This model features 60 titanium microelectrodes that are 30 μm diameter and an interelectrode distance of 500 μm, distributed in a 6 × 10 array. Finally, the insulation layer of this substrate is silicon nitride.

3. Materials and Methods

For the analysis of the biocompatibility of the materials tested, mouse retinal explants were incubated within both chambers for 48 h. Different tests have been carried out to analyze the biocompatibility using tissue from young and adult mice. Animal care and experimentation followed the Association for Research in Vision and Ophthalmology Statement for the Use of Animals in Ophthalmic and Vision Research and the guidelines of the European Union Directive 2010/63/EU on the protection of animals used for scientific purposes and was approved by Institutional Animal Ethics Committee.

Regarding the retinal explants, the study focuses on the neuroretina, a multilayered tissue composed of different types of neurons and glial cells. The photoreceptors, which are the photosensitive neurons, constitute the outmost layer, and the ganglion cells that transmit the information to the brain are in the innermost layer. Retinal explants were prepared from C57BL/6 wild-type mice of ages ranging from 14 days post-natal to 60 days (adult). A standard aseptic dissection technique [31] was used, in which retinal pigment epithelium was peeled off from the neuroretina, and each of the neuroretinas was flattened from their original spherical shape by four scissor cuts, rendering a "flower" shape known as flat mount. To avoid cellular damage induced by excessive manipulation and to ease the placement of the tissue in the system, the dissected neuroretinas were mounted onto a perforated polyvinylidene fluoride (PVDF) frame. Neuroretinas were arranged in each of the systems with the photoreceptor layer facing toward the material being tested for biocompatibility, with the ganglion cell layer facing upward. Interphase with the culture media was kept through a permeable Millicell (Millipore PIOCM0RG50) membrane on top pf the PVDF/retina structure. The biological material was fed with culture media (Neurobasal-A (Gibco 10888022), 2% B27 (Gibco 7504044), 1% N2 (Gibco 17502048), 1% pen/strp (Gibco 15140122), and 0.4% Glutamax (Gibco 35050038)).

In a first approach to measure the cell viability of retinas cultured onto the tested materials, a static/open system was set using 14-days postnatal retinas. In this case, taking into account the fact that the system is a static/open mount, the more adequate control is to maintain mouse retinas onto a standard Millicell cell culture insert inside a regular cell culture plate. However, this is not a proper control for a closed/continuous flow system, since the cultures' conditions are too different to be compared. For comparative purposes for the test in closed/continuous flow, we have used the commercial MEA of MultiChannel. A microfluidic platform is manufactured in a 3D printer and glued to the MEA. For the biocompatibility test of this setting, the only material to be studied is the 3D resin, since the rest of materials (Silicon Nitride, PVDF, and Millicell) are known to be biocompatible. Moreover, the resin is only in contact with the culture media and not with the tissue, so the biocompatibility experiment has to replicate this situation, and in this case it will not be relevant to test the biocompatibility by culturing the retinas in direct contact with the material. The three different open culture systems can be seen in Figure 2.

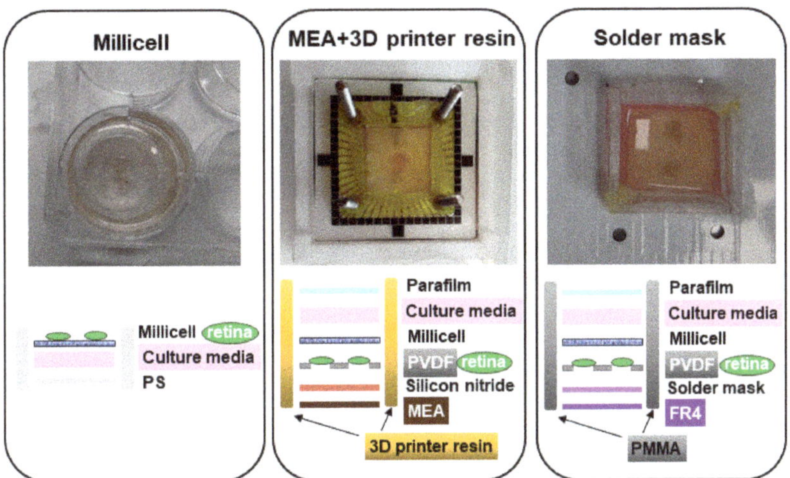

Figure 2. Images of the three different open culture systems and schematic representation of the setting for each case.

Explants were laid onto the PCB, completely covered with solder mask, (with a culture chamber made of PMMA), and laid into a resin-based culture chamber (attached to a biocompatible MEA as culture substrate), covered by 1 mL of culture media and maintained into a standard cell culture incubator at 5% CO_2, 20% O_2 atmosphere and 37 °C for 48 h. On top, a sterile sheet of Parafilm was used to avoid evaporation of the culture media. A standard Millicell culture insert was used as a control for cellular viability of the retinal explants. Once the culture time was finished, each retina was carefully retrieved from its culture system. In this experiment, dead retinal cells were stained using ethidium homodimer-1 and counterstained with calcein (live/dead viability/cytotoxicity kit for mammalian cells, molecular probes MP 03224) by incubating the tissue in 0.4% ethidium homodimer, 0.125% calcein in D-PBS (Merk 59331CC) for 30 min at 37 °C before proceeding to tissue disaggregation to obtain single cells. Retinas were incubated for 5 min with 100 μL of a mixture of proteolytic and collagenolytic enzymes (Accutase, Stemcell 07920) at 37 °C, followed by soft agitation using 10 up-down pipet cycles with a 1 mL micropipette, and 100 μL of a 1 mg/mL DNase I (Roche 10104159001) solution in D-PBS (Merk 59331C) was added and incubated for additional 5 min at 37 °C. Soft agitation was performed again five times, and the cell suspension was filtered through a 40 μL cell strainer (Falcon 352340). Cytometry buffer (2% FBS (ThermoFisher Scientific 16140071), 1 mM EDTA (Sigma E5134) in D-PBS) was added to complete 1 mL. Biocompatibility was tested via flow cytometry using a FACSCalibur flow cytometer (BD Biosciences, San Jose, CA, USA) equipped with 488 nm and 635 nm lasers. Analysis of the data was carried out using FlowJo v10.7 software (BD Biosciences, San Jose, CA, USA) and will be discussed in Section 4.

In order to evaluate biocompatibility of PCB and resin separately, as stand-alone culture systems independent from a cell incubator, a closed system was set for the PCB-PMMA LoP and for the MEA-resin lab on chip (LoC) systems (shown in Figure 3). A continuous flow of cell culture media was sustained by a syringe pump (new era pump systems NE-1000) at a 3.6 μL/min rate and maintained at a 37 °C temperature in a temperature-controlled chamber. Culture media was equilibrated in 5% CO_2, 20% O_2 atmosphere before starting explant feeding. Adult mouse retinas and a different staining technique were used in this occasion. Tissue was disaggregated before staining, and dead cells were labeled with 2-(4-amidinophenyl)-1H-indole-6-carboxamidine (DAPI; 1:2500 in cytometry buffer) and measured by flow cytometry. DAPI is a fluorescent molecule that binds strongly to DNA but cannot pass through an intact cell membrane and, therefore, preferentially stains dead cells.

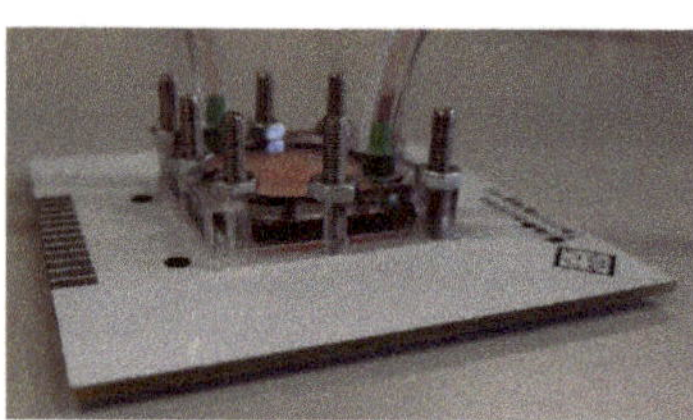
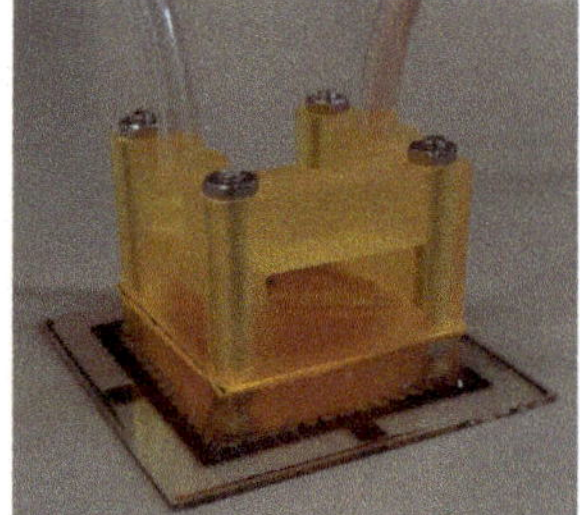

Figure 3. The closed systems evaluated are shown: on the right side, the commercial MEA + 3D printer resin, and on the left, the PCB system.

Taking into account that our main focus is the preparation of LoC systems aimed to study retinal degeneration and to test possible means of neuroprotection, we are specifically interested in the retinal cells that degenerate in our mouse model, which are the rod photoreceptor cells. Rod cells express rhodopsin. To determine the preservation of the rod cells after explant culture in these systems, we used cells from the closed system experimental conditions and labeled them with an antibody against rhodopsin.

For the flow cytometry analysis of rhodopsin-expressing cells, single cells were incubated for 30 min with Violet Live/Dead fixable stain (ThermoFisher Scientific, Waltham, MA, USA) and fixed with BD Cytofix/Cytoperm^{TM} kit (BD Biosciences, San Jose, CA, USA). After fixation and permeabilization, cells were stained with a primary antibody against rhodopsin (Abcam # ab190307, 1:100) for 30 min on ice, followed by washes and staining with donkey anti-mouse Alexa Fluor 488 (ThermoFisher #A-21202, 1:1000) secondary antibody for another 30 min on ice. Fluorescence minus one (FMO) controls were included for each condition to identify gate-negative and -positive cells. Stained cells were analyzed using a LSRFortessa X-20 flow cytometer equipped with 488 nm, 561 nm, 405 nm, and 640 nm lasers (BD Biosciences, San Jose, CA, USA). Analysis of the data was carried out using FlowJo v10.7 software (BD Biosciences, San Jose, CA, USA). Only single cells data was considered for the analysis, excluding cell fragments and cell aggregates.

4. Experimental Results

For the analysis of the biocompatibility of the solder mask with retinal explants, mouse retinas were prepared with a standard explant technique and cultured for two days in contact with this material. As previously commented, we first used an open configuration of the culture system, and organotypic culture was performed into a standard cell culture incubator. A regular Millicell retinal explant culture was used as control, in the first test, to compare cell viability. Additionally, a MEA + 3D printer resin was also studied, for comparative purposes in the following experiments in closed configuration. Data of Figures 4 and 5 correspond to 10,000 recorded cells from each of two retinas in each condition.

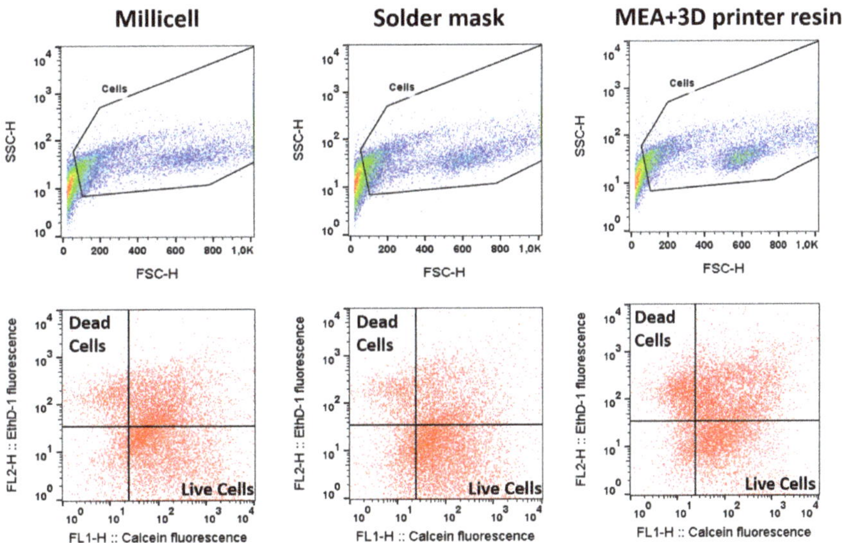

Figure 4. Biocompatibility measured in open culture systems. Representative images of cytometer acquisition data of cells cultured on Millicell insert, on solder mask, and on MEA + 3D printer resin, as shown from left to right. The upper row shows the gating of single cells. The lower row shows calcein fluorescence in the x-axis and ethidium fluorescence in the y-axis, corresponding to live and dead cells, respectively.

Flow cytometry analysis of the retinal cells stained with Ethidium homodimer-1 and calcein A shows that the solder mask preserves retinal cell viability in a similar proportion to the standard Millicell insert used as a control. We found that retinal explants cultured in the MEA + 3D printer resin system present a slightly lower cell preservation of cell viability. However, we considered that this difference in cell viability was small enough to still consider the MEA + 3D printer resin system a potentially good experimental setting for subsequent testing. Considering our aim and future experimental purposes, both PCB-solder mask and MEA + 3D-printer resin sustain the culture of retinal explants adequately to test neuroprotective agents and to study the degenerative process in these culture conditions. Altogether, these results indicate that neither the solder mask nor the resin exert a cytotoxic effect on cell viability, and that both materials displayed acceptable levels of biocompatibility.

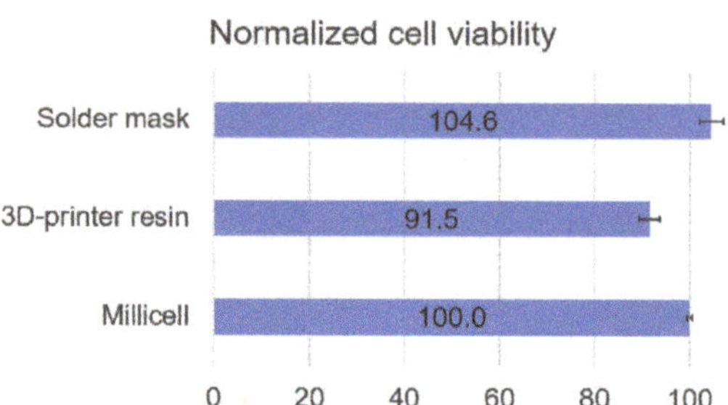

Figure 5. Biocompatibility measured in open culture systems. Bar plot displaying the viability of retinal cells after being cultured with three different materials for 48 h in open systems. Percentage of viable cells after culture of retinal explants onto Millicell membrane, PCB, and MEA + 3D printer resin. The error bars represent the standard deviation.

To replicate this test in an autonomous LoC setting, the retinal explants were cultured for 48 h in a closed configuration outside a cell culture incubator (Figure 3), and the biological material was fed via continuous pumping of culture media. Then, we used flow cytometry to compare the viability of retinas cultured in the PCB-solder mask system (in direct contact with the solder mask) to the viability of retinas cultured on a closed system composed by a commercial MEA and a microfluidic platform made of 3D printer resin (where retinas are in direct contact with the silicon nitride layer of the MEA). Data of Figures 6 and 7 correspond to 10,000 recorded cells from each of three retinas for each condition.

Preservation of the biological material was shown even using adult retinas, which usually present a lower viability in explant culture. In concordance with the results of the previous experimental setting, cell viability preservation was found to be slightly higher when retinal explants were cultured in contact with the solder mask compared to the MEA-3D-printer resin set up.

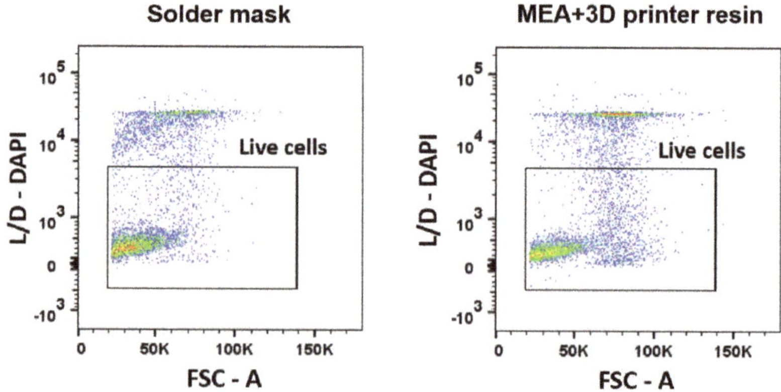

Figure 6. Biocompatibility measured in closed culture systems. Representative dot plots of cytometer acquisition data of cells cultured in PCB in a closed system and resin-based LoC system. Dead cells are positively stained for DAPI, which is displayed in the Y-axis.

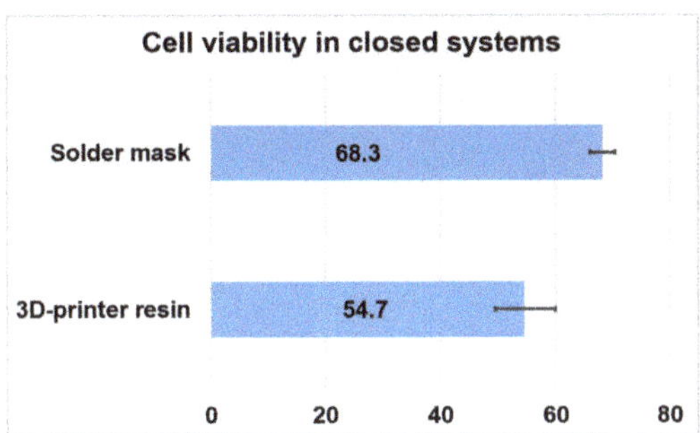

Figure 7. Biocompatibility measured in closed culture systems. Bar plot displaying the viability of retinal cells after being cultured with two different materials for 48 h in closed systems. Percentage of viable cells after culture of retinal explants in autonomous systems onto PCB, and onto a resin-based LoC. The error bars represent the standard deviation.

The long-term goal of this study is the development of a functional LoC system for testing the effect of neuroprotective agents on photoreceptor degeneration occurring in model systems of retinitis pigmentosa, such as the rd10 mice. Therefore, we specifically tested the viability of rod cells in the aforementioned closed culture systems. The specific cell population affected in the rd10 mice model is rod photoreceptor cells, which express rhodopsin. To determine the preservation of the rod cells after organotypic culture into this system, we quantified the number of rhodopsin-positive cells after fixing and staining them with a rhodopsin antibody.

Our results (Figure 8) show good preservation of rod cells that is similar in both conditions tested, with 97.1% of rhodopsin positive events for the cells cultured onto PCB and 88.9% for the resin-based device. Therefore, our data indicate that both the PCB and the resin have no specific toxicity for the rod photoreceptor cells, which were in contact with these materials for 48 h. In line with our previous results, PCB-solder mask presents a better preservation of retinal cells—in this case, specifically rod photoreceptors—than 3D printer resin. However, the yet good rod photoreceptor conservation observed in the MEA-3D printer resin set up (>88%) demonstrates that this system can still be considered a good candidate for organotypic retinal culture.

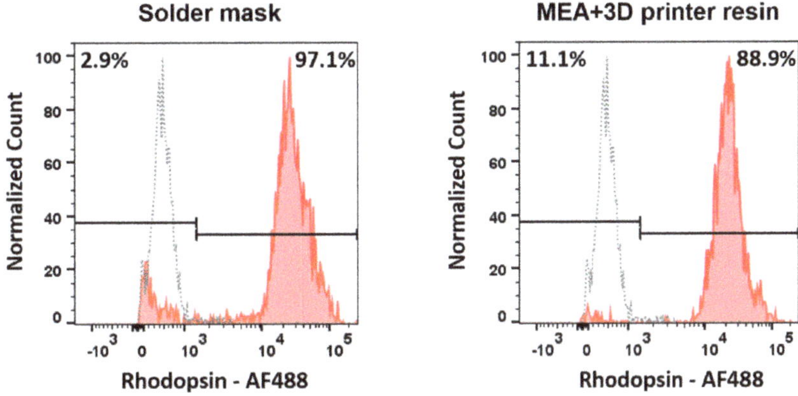

Figure 8. Histograms displaying the percentage of rhodopsin positive cells after culturing retinal explants onto PCB system vs. commercial MEA and 3D printer resin. Fluorescence minus one control (FMO) is displayed in gray as a negative control used for setting the quantification gates.

5. Conclusions

Different experiments of cell viability have been performed to evaluate the biocompatibility of the solder mask of the PCB. Moreover, the biocompatibility of a printable resin has also been carried out. Both materials were in contact with the retinal explant culture, and the results show both of them are compatible with retinal cultures.

Our data support the use of PCB (covered with this solder mask) as a component of organotypic culture systems and, more specifically, in systems designed to evaluate retinal explants for the study of photoreceptor degeneration and neuroprotection. In Lab-on-PCB, where the integration of microfluidics and electronics is required, this is an essential point. Therefore, these results sustain the future target of fabricating the MEA on PCB, for studies on regenerative medicine to develop methods to improve vision. However, if the dimension requirements about electrodes and tracks for a specific application are more demanding than the commercial PCB manufacturer offers, the MEA of MultiChannel Systems, integrated to a microfluidic platform made of 3D printer resin, could serve as an alternative, although at a higher economic cost. Another significant advantage is that it could be considered an open-source design Klayout and FreeCad software are used for the design. A stand-alone Lab-on-PCB system in which the retinal explants can be constantly monitored for the effect of different stressors or neuroprotective agents/conditions would provide valuable information on the neurodegenerative process that leads to cell death and sight loss. It would also provide a tool to test different approaches against cell death caused by retinal degenerative diseases. Specifically, we plan to use these devices to test the effect of different doses of electrostimulation, the different methods of gene therapy to address degenerating cells in the retina, and new agents with neuroprotective properties that will be added to the circulating culture media. Once the utility of these systems is established using retinal explants, we plan to apply this technology to the evaluation of retinal organoids derived by in vitro differentiation from cells of human healthy donors and patients, avoiding the use of laboratory animals and obtaining information closer to the real human patient. The use of the Lab-on-PCB would not be restricted to research in the retinal field; it could be expanded to wider biomedical applications, culturing different cell types, other organotypic cultures, or patient-derived organoids.

The possibility of developing low-cost MEAs, customized to other biomedical applications, with a cost about 10 times cheaper than commercial MEAs with similar characteristics, will offer the researchers great flexibility for designing their own MEAs and experimental settings.

Author Contributions: Conceptualization, F.P., B.d.l.C. and C.A.; methodology, J.D.U.-G., B.d.l.C. and L.V.-S.; software, Á.P.R.; validation, I.R.L.; formal analysis, B.d.l.C., L.V.-S. and Á.P.R.; investigation, F.P. and C.A.; resources, B.d.l.C. and J.D.U.-G.; data curation, F.P.; writing—original draft preparation, F.P., B.d.l.C. and J.D.U.-G.; writing—review and editing, C.A. and J.D.U.-G.; visualization, all of authors; supervision, F.J.D.-C.; project administration, C.A.; funding acquisition, C.A. and J.M.Q. All authors have read and agreed to the published version of the manuscript.

Funding: This research was funded by Fondo Europeo de Desarrollo Regional (FEDER) and Consejería de Economía, Conocimiento, Empresas y Universidad de la Junta de Andalucía (Programa Operativo FEDER 2014–2020), Lab-on-chip de electro-estimulación, para el estudio In-vitro de Cultivos de retina de Larga duración: Retina-on-a-chip project number US-1265983 (Universidad de Sevilla).

Institutional Review Board Statement: The study was conducted according to the guidelines of the Declaration of Helsinki, and approved by the Institutional Ethics Committee of CABIMER, CEEA-CABIMER, with the protocol code 04/04/2019/058 and date of approval 25 March 2019).

Informed Consent Statement: Not applicable.

Data Availability Statement: Data is contained within the article.

Conflicts of Interest: The authors declare no conflict of interest.

Abbreviations

The following abbreviations are used in this manuscript:

PCB	Printed Circuit Board
MEA	Microelectrode array
LoP	Lab-on-PCB
PMMA	Polymethylmethacrylate
EVA	Ethylene-vinyl acetate
LoC	Lab-on-chip
PVDF	Polyvinylidene fluoride
PS	Polystyrene
FMO	Fluorescence minus one

References

1. Merkel, T.; Graeber, M.; Pagel, L. A new technology for fluidic microsystems based on PCB technology. *Sens. Actuators A Phys.* **1999**, *77*, 98–105. [CrossRef]
2. Zupančič, U.; Rainbow, J.; Estrela, P.; Moschou, D. Utilising Commercially Fabricated Printed Circuit Boards as an Electrochemical Biosensing Platform. *Micromachines* **2021**, *12*, 793. [CrossRef] [PubMed]
3. Zhao, W.; Tian, S.; Huang, L.; Liu, K.; Dong, L. The review of Lab-on-PCB for biomedical application. *Electrophoresis* **2020**, *41*, 1433–1445. [CrossRef]
4. Perdigones, F. Lab-on-PCB and Flow Driving: A Critical Review. *Micromachines* **2021**, *12*, 175. [CrossRef]
5. Nikshoar, M.S.; Khosravi, S.; Jahangiri, M.; Zandi, A.; Miripour, Z.S.; Bonakdar, S.; Abdolahad, M. Distinguishment of populated metastatic cancer cells from primary ones based on their invasion to endothelial barrier by biosensor arrays fabricated on nanoroughened poly (methyl methacrylate). *Biosens. Bioelectron.* **2018**, *118*, 51–57. [CrossRef] [PubMed]
6. Jolly, P.; Rainbow, J.; Regoutz, A.; Estrela, P.; Moschou, D. A PNA-based Lab-on-PCB diagnostic platform for rapid and high sensitivity DNA quantification. *Biosens. Bioelectron.* **2019**, *123*, 244–250. [CrossRef]
7. Kaprou, G.D.; Papadopoulos, V.; Loukas, C.M.; Kokkoris, G.; Tserepi, A. Towards PCB-based miniaturized thermocyclers for DNA amplification. *Micromachines* **2020**, *11*, 258. [CrossRef]
8. Kaprou, G.; Papadakis, G.; Papageorgiou, D.; Kokkoris, G.; Papadopoulos, V.; Kefala, I.; Gizeli, E.; Tserepi, A. Miniaturized devices for isothermal DNA amplification addressing DNA diagnostics. *Microsyst. Technol.* **2016**, *22*, 1529–1534. [CrossRef]
9. Anastasova, S.; Kassanos, P.; Yang, G.Z. Multi-parametric rigid and flexible, low-cost, disposable sensing platforms for biomedical applications. *Biosens. Bioelectron.* **2018**, *102*, 668–675. [CrossRef]
10. Gautam, V.; Rand, D.; Hanein, Y.; Narayan, K. A polymer optoelectronic interface provides visual cues to a blind retina. *Adv. Mater.* **2014**, *26*, 1751–1756. [CrossRef]
11. Multichannel Systems. Available online: https://www.multichannelsystems.com/ (accessed on 16 August 2021).
12. Luk, C.C.; Lee, A.J.; Wijdenes, P.; Zaidi, W.; Leung, A.; Wong, N.Y.; Andrews, J.; Syed, N.I. Trophic factor-induced activity 'signature'regulates the functional expression of postsynaptic excitatory acetylcholine receptors required for synaptogenesis. *Sci. Rep.* **2015**, *5*, 9523. [CrossRef]
13. Urbano-Gámez, J.D.; Casañas, J.J.; Benito, I.; Montesinos, M.L. Prenatal treatment with rapamycin restores enhanced hippocampal mGluR-LTD and mushroom spine size in a Down's syndrome mouse model. *Mol. Brain* **2021**, *14*, 84. [CrossRef]
14. Hermann, D.; van Amsterdam, C. Analysis of spontaneous hippocampal activity allows sensitive detection of acetylcholine-mediated effects. *J. Pharmacol. Toxicol. Methods* **2015**, *71*, 54–60. [CrossRef] [PubMed]
15. Belle, M.D.; Baño-Otalora, B.; Piggins, H.D. Perforated multi-electrode array recording in hypothalamic brain slices. In *Circadian Clocks*; Humana Press: New York, NY, USA, 2021; pp. 263–285.
16. Fu, F.; Chauhan, M.; Sadleir, R. The effect of potassium chloride on Aplysia Californica abdominal ganglion activity. *Biomed. Phys. Eng. Express* **2018**, *4*, 035033. [CrossRef]
17. Trantidou, T.; Terracciano, C.M.; Kontziampasis, D.; Humphrey, E.J.; Prodromakis, T. Biorealistic cardiac cell culture platforms with integrated monitoring of extracellular action potentials. *Sci. Rep.* **2015**, *5*, 11067. [CrossRef] [PubMed]
18. Cabello, M.; Aracil, C.; Perdigones, F.; Quero, J.M.; Rocha, P.R. Lab-on-pcb: Low cost 3d microelectrode array device for extracellular recordings. In Proceedings of the 2018 Spanish Conference on Electron Devices (CDE), Salamanca, Spain, 14–16 November 2018; pp. 1–4.
19. Cabello, M.; Mozo, M.; De la Cerda, B.; Aracil, C.; Diaz-Corrales, F.J.; Perdigones, F.; Valdes-Sanchez, L.; Relimpio, I.; Bhattacharya, S.S.; Quero, J.M. Electrostimulation in an autonomous culture lab-on-chip provides neuroprotection of a retinal explant from a retinitis pigmentosa mouse-model. *Sens. Actuators B Chem.* **2019**, *288*, 337–346. [CrossRef]
20. Constantin, C.P.; Aflori, M.; Damian, R.F.; Rusu, R.D. Biocompatibility of polyimides: A mini-review. *Materials* **2019**, *12*, 3166. [CrossRef] [PubMed]
21. Insanally, M.; Trumpis, M.; Wang, C.; Chiang, C.H.; Woods, V.; Palopoli-Trojani, K.; Bossi, S.; Froemke, R.C.; Viventi, J. A low-cost, multiplexed μECoG system for high-density recordings in freely moving rodents. *J. Neural Eng.* **2016**, *13*, 026030. [CrossRef] [PubMed]

22. JLCPCB. Available online: https://support.jlcpcb.com/article/64-certifications (accessed on 16 August 2021).
23. Ogilvie, J.M.; Speck, J.D.; Lett, J.M.; Fleming, T.T. A reliable method for organ culture of neonatal mouse retina with long-term survival. *J. Neurosci. Methods* **1999**, *87*, 57–65. [CrossRef]
24. Müller, B.; Wagner, F.; Lorenz, B.; Stieger, K. Organotypic cultures of adult mouse retina: Morphologic changes and gene expression. *Investig. Ophthalmol. Vis. Sci.* **2017**, *58*, 1930–1940. [CrossRef]
25. Ou, Y.T.; Lu, M.S.C.; Chiao, C.C. The effects of electrical stimulation on neurite outgrowth of goldfish retinal explants. *Brain Res.* **2012**, *1480*, 22–29. [CrossRef] [PubMed]
26. Hughes, S.; Rodgers, J.; Hickey, D.; Foster, R.G.; Peirson, S.N.; Hankins, M.W. Characterisation of light responses in the retina of mice lacking principle components of rod, cone and melanopsin phototransduction signalling pathways. *Sci. Rep.* **2016**, *6*, 28086. [CrossRef] [PubMed]
27. Tsai, D.; Sawyer, D.; Bradd, A.; Yuste, R.; Shepard, K.L. A very large-scale microelectrode array for cellular-resolution electrophysiology. *Nat. Commun.* **2017**, *8*, 1802. [CrossRef] [PubMed]
28. Tu, H.Y.; Watanabe, T.; Shirai, H.; Yamasaki, S.; Kinoshita, M.; Matsushita, K.; Hashiguchi, T.; Onoe, H.; Matsuyama, T.; Kuwahara, A.; et al. Medium-to long-term survival and functional examination of human iPSC-derived retinas in rat and primate models of retinal degeneration. *EBioMedicine* **2019**, *39*, 562–574. [CrossRef] [PubMed]
29. Cowan, C.S.; Renner, M.; De Gennaro, M.; Gross-Scherf, B.; Goldblum, D.; Hou, Y.; Munz, M.; Rodrigues, T.M.; Krol, J.; Szikra, T.; et al. Cell types of the human retina and its organoids at single-cell resolution. *Cell* **2020**, *182*, 1623–1640. [CrossRef] [PubMed]
30. Urbano-Gámez, J.D.; Aracil, C.; Perdigones, F.; Fontanilla, J.A.; Quero, J.M. Towards a 3D-Printed and Autonomous Culture Platform Integrated with Commercial Microelectrode Arrays. In Proceedings of the 2021 13th Spanish Conference on Electron Devices (CDE), Sevilla, Spain, 9–11 June 2021; pp. 153–155.
31. Müller, B. *Organotypic culture of adult mouse retina. Mouse Cell Culture*; Humana Press: New York, NY, USA, 2019; pp. 181–191.

micromachines

MDPI

Article

A Novel Planar Grounded Capacitively Coupled Contactless Conductivity Detector for Microchip Electrophoresis

Jianjiao Wang [1], Yaping Liu [1], Wenhe He [2], Yuanfen Chen [1] and Hui You [1,*]

[1] School of Mechanical Engineering, Guangxi University, Nanning 10593, China; 1811301028@st.gxu.edu.cn (J.W.); 2011391061@st.gxu.edu.cn (Y.L.); yuanfenchen@gxu.edu.cn (Y.C.)
[2] School of Electrical Engineering, Guangxi University, Nanning 10593, China; 1812401002@st.gxu.edu.cn
* Correspondence: usmlhy@iim.ac.cn

Abstract: In the microchip electrophoresis with capacitively coupled contactless conductivity detection, the stray capacitance of the detector causes high background noise, which seriously affects the sensitivity and stability of the detection system. To reduce the effect, a novel design of planar grounded capacitively coupled contactless conductivity detector (PG-C4D) based on printed circuit board (PCB) is proposed. The entire circuit plane except the sensing electrodes is covered by the ground electrode, greatly reducing the stray capacitance. The efficacy of the design has been verified by the electrical field simulation and the electrophoresis detection experiments of inorganic ions. The baseline intensity of the PG-C4D was less than 1/6 of that of the traditional C4D. The PG-C4D with the new design also demonstrated a good repeatability of migration time, peak area, and peak height (n = 5, relative standard deviation, RSD $\leq$ 0.3%, 3%, and 4%, respectively), and good linear coefficients within the range of 0.05–0.75 mM ($R^2 \geq 0.986$). The detection sensitivity of K^+, Na^+, and Li^+ reached 0.05, 0.1, and 0.1 mM respectively. Those results prove that the new design is an effective and economical approach which can improve sensitivity and repeatability of a PCB based PG-C4D, which indicate a great application potential in agricultural and environmental monitoring.

Keywords: microchip electrophoresis; stray capacitance; capacitively coupled contactless conductivity detection; planar grounded electrode

Citation: Wang, J.; Liu, Y.; He, W.; Chen, Y.; You, H. A Novel Planar Grounded Capacitively Coupled Contactless Conductivity Detector for Microchip Electrophoresis. *Micromachines* **2022**, *13*, 394. https://doi.org/10.3390/mi13030394

Academic Editor: Antonio Ramos

Received: 30 January 2022
Accepted: 25 February 2022
Published: 28 February 2022

1. Introduction

Microchip electrophoresis has the characteristics of small size (~cm^2) and short separation channels, which is efficient and convenient for on-site and real-time detection. Combined with C4D, microchip electrophoresis device is easy to miniaturize, avoiding solution contamination, as well as large and precise instrument requirement. Thus, C4D technology has been widely applied in capillary electrophoresis and microchip electrophoresis [1–4].

C4D was first proposed independently by Zemann et al. [5] and da Silva et al. [6], applied in capillary electrophoresis. Subsequently, Pumera and Wang [7] replaced the tubular electrode in C4D with two flat aluminum sheets and applied it in microchip electrophoresis. When the sample passed through the detection cell along the channel, the change of the local impedance caused the change of the alternating current on the pick-up electrode. After rectification, filtering, and amplification, the current on pick-up electrode was sampled and displayed as peaks [8,9].

At present, the common problems in microchip electrophoresis with C4D are high baseline background intensity, serious noise interference, poor detection sensitivity, and repeatability. These problems are mainly caused by stray capacitance and wall capacitance [10–12]. The wall capacitance can be reduced by integrated designs to control the thickness and material of the insulating layer [13–15]. Moreover, reasonably increasing the excitation frequency can also reduce the influence of wall capacitance. However, integrated designs are usually slightly more complex and increase the cost of the system. Meanwhile, increasing the operating frequency will increase the interference caused by stray capacitance. In severe cases, the interference even dominates the detector output, which has a fatal impact on the detection performance.

The method of arranging a ground electrode between the sensing electrodes to reduce the stray capacitance has been widely applied [8,16–19]. However, it is generally considered that its effect on reducing stray capacitance is limited. Based on the idea of ground electrode, Guijt et al. [20] built an integrated device design on a PCB to study the effect of insulation thickness on detection performance. Mahanadi et al. [21] added a pair of sensing electrodes to the traditional structure, forming a dual top-bottom geometry. This method doubled the coupling efficiency of the electrodes. However, the designs of microchip and electrodes are complicated, which may be the reason this method has not been adopted widely.

The other way is to counteract the adverse effects of stray capacitance and wall capacitance according to differential mode and impedance cancellation (circuit resonance) principle. Based on the principle of impedance cancellation, Shen et al. [22,23] obtained the minimum impedance with a low-impedance capacitively coupled contactless conductivity detector (LIC4D) at the resonance frequency. This method is achieved by connecting a piezoelectric quartz crystal (PQC, acts as an inductor) in series between the C4D cell and the excitation signal. Compared with traditional C4Ds, the signal-to-noise ratio of LIC4D was increased by more than 20 times. One problem is that the operating frequency of the actual inductor was difficult to adjust [24]. Based on differential mode, Laugere [25] introduced an integrated four-electrode design of the detector. Compared with the traditional bipolar configuration, the reported detector had a better signal-to-noise ratio in a wide frequency range, as low as 600 Hz. Georg Fercher [26] also adopted the similar method, and the obtained experimental result showed that the baseline intensity was greatly reduced, and the detection sensitivity was increased by several times. Xiao [27] and Stojkovic M et al. [28] introduced differential mode into capillary electrophoresis, which also improved the sensitivity of the system. Although many researchers have proposed solutions to counteract the adverse effects of stray capacitance, complex design requires complicated and strict fabrication processes and increases the cost of the system. An economical, easy-to-fabricate, sensitive C4D design for microchip electrophoresis is in demand, especially for the large-scale applications in environmental monitoring.

In this study, we described a planar grounded C4D (PG-C4D) based on PCB for microchip electrophoresis. The electrophoretic responses demonstrated that the PG-C4D had smaller background baseline intensity than traditional C4D, which can be attributed to the reduction in stray capacitance. Subsequently, the electrophoretic detection of three inorganic cations, K^+, Na^+, and Li^+, proved that PG-C4D had better signal-to-noise ratio and repeatability. This simple manufactured, low-cost, and sensitive PG-C4D for microchip electrophoresis has great potential application in field applications.

2. Materials and Methods

2.1. Prototyping of Microfluidic Devices

The layout of the microchip device with PCB based PG-C4D is shown in Figure 1. The microchip was secured on the PCB based C4D using a clamp and two screws, as shown in Figure 1A. The C4D electrodes were fabricated on a commercially available PCB (Shenzhen JLC Electronics Co., Ltd., Shenzhen, China). The solder mask was removed to reduce the distance between the electrodes and the microchip. The C4D electrodes were composed of a pair of sensing electrodes (1 mm width, 2 mm length, and 35 μm thickness) and a ground electrode. The sensing electrodes were spaced by an 800 μm gap, optimized in the previous reported work from our group [29]. The difference between the PG-C4D and the traditional C4D was the ground electrode design. The ground electrode in PG-C4D surrounded the sensing electrodes, while the ground electrode in traditional C4D lay in the middle of the sensing electrode pairs with width of 200 μm, as shown in Figure 1B,C.

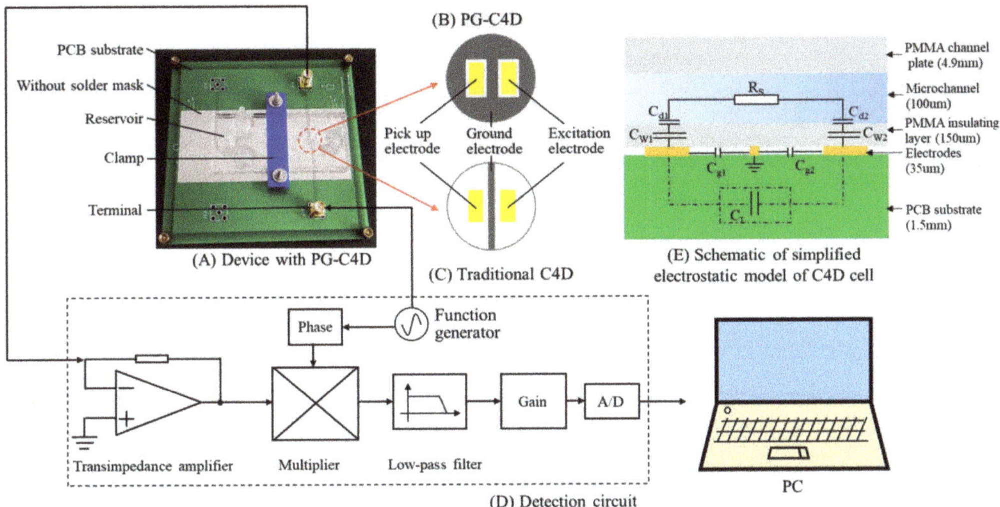

Figure 1. Schematic drawing of a microchip device with external C4D. (**A**) Device with PG-C4D (**B**) PG-C4D. (**C**) Traditional C4D. (**D**) Detection circuit. (**E**) Schematic of simplified electrostatic model of C4D cell.

The microchips consisted of a 5 mm thick PMMA plate and a 150 μm thick PMMA film. PMMA plate was used as the microchannel layer. The PMMA film was used as the insulating layer. The microchip was fabricated using the following steps. First, the microchannels in the PMMA plate were fabricated by a CNC machine (X5 combo, SYIL MACHINE TOOLS Co., Ltd., Ningbo, China). The length of the injection and separation microchannels were 16 mm and 53 mm, respectively. Both microchannels had a square cross section of 100 μm × 100 μm. Secondly, the PMMA plate with microchannels was cleaned with purified water in an ultrasonic cleaner (SB-5200DT, NINGBO SCIENTZ BIOTECHNOLOGY Co., Ltd., Ningbo, China) three times, 10 min each time. Then, the PMMA plate along with the PMMA film were air plasma treated at 700 mTorr at high power (30 W) for 5 min (PDC-002, Harrick Plasma, Ithaca, NY, USA). Next, the PMMA plate with microchannels was aligned onto the PMMA film and thermally pressed at 105 °C under 0.3 MPa for 15 min, leading to a strong adhesion between the layers. Finally, the bases of plastic pipette tips were attached on the top of reservoirs on the PMMA plate, in order to connect the microchip with external devices. A device of microchip electrophoresis with PG-C4D based on PCB, CNC and thermal press packaging is easy to process and economical, which is favorable for commercialization.

2.2. Reagents and Electrophoretic Procedures

All reagents were of analytical grade and purchased from Macklin (Shanghai, China). The solutions were prepared using deionized water (resistivity 18.2 MΩ · cm) processed through a purification system (Cascada I, PALL, Beijing, China). Stock solutions of cations (K^+, Na^+, and Li^+, 5 mM) were prepared from their corresponding chloride salts. The 60 mM stock solutions of L-histidine (His) and 2-(N-morpholino)-ethanesulfonic acid (MES) were prepared daily. A mixture of 20 mM MES and 20 mM His at pH 6.0 was used as running buffer, and was filtered through 0.22 um nylon syringe filters before use.

To improve the repeatability, prior to use, the microchannels were washed with 100 mM sodium hydroxide solution, deionized water, and running buffer for 15 min, respectively. The sample solution was prepared daily in running buffer, avoiding sample stacking. At the end of a working day, the microchips were rinsed with deionized water, to prevent clogging and contamination. Sample injection was performed electrokinetically using cross-injection method.

When the sample passed through the detection cell, the cell impedance would change, thus the sinusoidal current output of the PG-C4D would change. The I/V conversion was completed by a transimpedance amplifier (OPA656, Texas Instruments, Austin, TX, USA) with a feedback resistor of 2MΩ [30]. This voltage signal was then lock-in amplified [31], filtered and gained, and finally sampled by an acquisition unit as shown in Figure 2. The high-voltage modules and associated circuits used in the system have been described in the previous work [32]. The data acquisition was done by a data acquisition unit (MAX194, Maxim Integrated, San Jose, CA, USA) and a software written in LABVIEW (National Instruments, Austin, TX, USA). The software was also used to control the high voltage switching from injection to separation step.

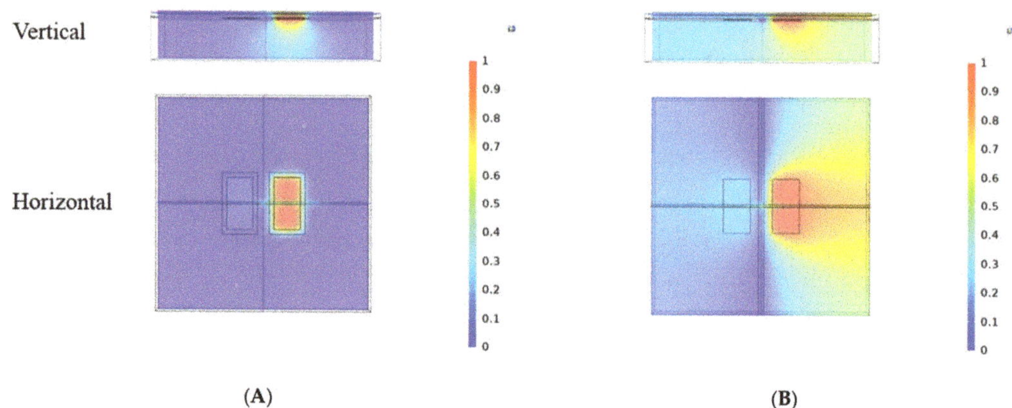

Figure 2. Simulated electric potential distribution of C4D cell, with both vertical and horizontal direction. A model (**A**) with PG-C4D feature and (**B**) with traditional C4D feature.

2.3. Electrostatic Model and Simulation Settings of C4D

In order to obtain the influence of stray capacitance on solution conductivity detection, the equivalent circuit model was analyzed using finite element simulation. C4D with different ground electrode design adopted the electrical model shown in Figure 1E, which was simplified from an RC network proposed by da Silva and do Lago et al. [5,6]. C_L is the stray capacitance. C_{W1} and C_{W2} are the wall capacitance of sensing electrodes coupled to the microchip, which are equal to C_W. C_{g1} and C_{g2} are the coupling capacitance between the sensing electrode and the ground electrode, which has negligible effect on the output current of the C4D [10]. The electric double layer C_{d1} and C_{d2} are very important in contact conductivity detection, but can be ignored in contactless conductivity detection [5]. Then the impedance of the C4D cell is defined as Z [11]. As shown in Equation (1), it is a function of C_W, R_S, and C_L.

$$Z = \frac{1}{j\omega C_L} // (R_S + \frac{2}{j\omega C_W}), \tag{1}$$

$$\frac{1}{Z} = 2j\pi f C_L + \frac{1}{\frac{1}{j\pi f C_W} + R_S}, \tag{2}$$

From Equation (1), R_S is connected in series with C_{W1} and C_{W2}, and then connected in parallel with C_L to form a cell impedance. The reciprocal of the impedance, as shown in Equation (2), shows more clearly how the C_L, C_W, and R_S affect the cell impedance. C_L will cause additional and unwanted signals, which are not affected by the change in conductivity of solution. The unwanted effect of C_L increases with frequency, and severely impairs the detection performance of the system.

To study the stray capacitance generated by C4D with different ground electrode designs, a cell model was established by the Multiphysics simulation software COMSOL 5.6. The microchannel section area and electrode dimensions were set as mentioned above. To improve calculation efficiency, only the space near the sensing electrode is simulated. The three-dimensional model of the prototyped PG-C4D is shown in Appendix A, Figure A1. System default material parameters were adopted. Detailed information of the C4D is listed in Table 1.

Table 1. Detailed information of the simulated C4D structure.

Title 1	Materials	Relative Permittivity	Size
Microchannel	Water	81	100 μm × 100 μm (cross-section)
PMMA insulating layer	PMMA	2.9	150 μm (thickness)
Sensing electrodes	Cu	1	2 mm × 1 mm × 35 μm (length × width × height)
PCB substrate	FR4	4.5	1.5 mm (thickness)

3. Results and Discussion

3.1. *Effect of Stray Capacitance on Cell Response*

The effect of stray capacitance in both PG-C4D and traditional C4D was simulated and calculated, demonstrating that planar grounded electrode would greatly decrease the stray capacitance. The simulation result was then validated by baseline strength testing.

The cell equivalent circuit and physical simulation model of C4D were established as mentioned in Section 2.3. The electric field distributions of both PG-C4D and traditional C4Ds were obtained. Figure 2A,B show the calculated electric field intensity distributions in vertical and horizontal direction for both C4Ds. For the PG-C4D, the electric field was concentrated around the electrode to which the positive potential was applied. This is due to the presence of the planar grounded electrode. While in traditional C4D, potential distributed away from the electrode of which the positive potential was applied. Furthermore, the potential of the other electrode without positive potential was not 0 V. The potential distribution of the traditional C4D meant that there was charge accumulation on the other sensing electrode (the terminal where no potential was applied). This indicates that there is stray capacitance between the sensing electrodes in traditional C4D. The electric field distribution indicated that traditional C4D had larger stray capacitance than PG-C4D.

The stray capacitance was calculated based on the simulation result. Detailed calculation steps are shown in Appendix B [33]. The stray capacitance of the PG-C4D was calculated to be 11.698fF, while the stray capacitance of traditional C4D was calculated to be 38.81fF, which was more than three times of that of the PG-C4D.

To confirm the simulation result that the stray capacitance of PG-C4D was greatly reduced, the output signal (the average of baseline intensity) of the transimpedance amplifier was tested. It can be seen from Figure 3A that, compared to traditional C4D, PG-C4D effectively reduced the baseline intensity to one order lower than that of traditional C4D. The measurement of peak to peak (output) can avoid interference from other parts of the circuit, which is an indicator of whether the stray capacitance had reduced or not, as shown in Figure 3B. The baseline intensity result was on par with the previous reported C4D with the ground plane perpendicular to the channel [16]. The layout of the PG-C4D was simpler, more convenient, and suitable for large-scale production, and showed a good suppression effect on stray capacitance. In addition, the output signal of the transimpedance amplifier became distorted once the excitation amplitude exceeded 10 V. As shown in diagram c of Figure 3B, this distorted phenomenon indicated that the signal cannot keep sinusoidal signal, due to the bandwidth and output voltage limitation of the transimpedance amplifier

(OPA656, ±5 V). This phenomenon also proved that the large stray capacitance in traditional C4D limited the applicable range of the excitation signal. The noise would also be more obvious under stronger excitation voltage. An effective method to introduce differential design would reduce the noise caused by high excitation voltage, but it would increase the cost of the system [20,26,28]. It could be concluded that PG-C4D can greatly reduce the interference current of the pick-up electrode due to the reduction of stray capacitance, and thus have better signal-to-noise ratio and detection sensitivity.

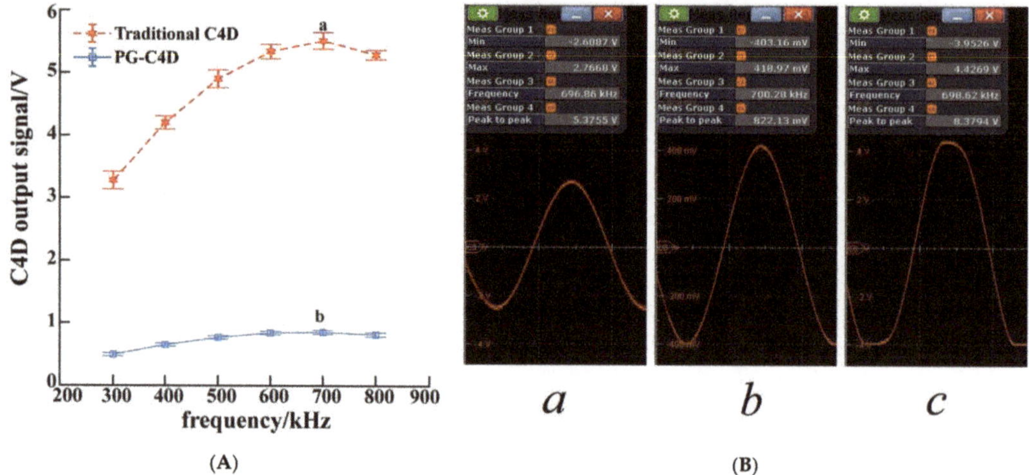

Figure 3. (**A**) C4D output signal. Red dashed line stands for the traditional C4D, and the blue continuous line stands for the PG-C4D. The error bars are the standard deviation of the measurements ($n = 3$). (**B**) Maximum output strength of the transimpedance amplifier. a: traditional C4D. b: PG-C4D. c: the excitation amplitude exceeded 10V in traditional C4D (distorted). Experimental conditions: channel filled with running buffer, oscilloscope filter cutoff frequency, 5 MHz.

3.2. Performance Comparison

The analytical performance of C4Ds with different ground electrodes was evaluated by inorganic cations detection experiment. The inorganic cations of potassium (K^+), sodium (Na^+), and lithium (Li^+) in the mixed solution of each ionic compounds were used as detection objects. The raw electropherograms of the three inorganic cations are shown in Figure 4. The RSD values of migration times, peak areas, and peak heights were calculated to compare the detection results of the two ground electrode configurations. Table 2 shows the calculated results for the 0.5 mM concentration level for five consecutive injections ($n = 5$).

The high-voltage switching at the moment of sampling would cause the output of electronic devices to jump, as shown by the arrows marked in Figure 4A,B. The inset diagrams in Figure 4 showed the raw baseline signal noise, without any filtering algorithm, to better estimate the ability of PG-C4D to suppress noise. It could be seen from the inset diagram in Figure 4A that the noise amplitude of the separated sample and running buffer measured by PG-C4D was approximately 4 mV. The peak heights for K^+, Na^+, and Li^+ were approximately 41.2 mV, 35.3 mV, and 23.7 mV, respectively. Here, the peak height was used to represent the peak intensity. The calculation process of the peak height was as follows. First, the signal was filtered, then peak function in MATLAB was applied to find the starting point, vertex, and end point of the peak. Finally, the peak height was obtained by subtracting the mean of the start and end points from the peak apex value. Signal-to-noise ratio of the peak height is greater than 3 for all the three ions. Though the concentrations of K^+, Na^+, and Li^+ were all the same in the sample, peak intensities decrease in the order of K^+, Na^+, and Li^+, which was related to the ionic conductivities of these ions [34]. The noise level in the traditional C4D is approximately 25 mV, as shown in the inset diagram in Figure 4B. The peak height for K^+, Na^+, and Li^+ were about 94.5 mV, 75.7 mV, and 54.7 mV, respectively. The noise in the signal seriously affects the performance of the device with traditional C4D.

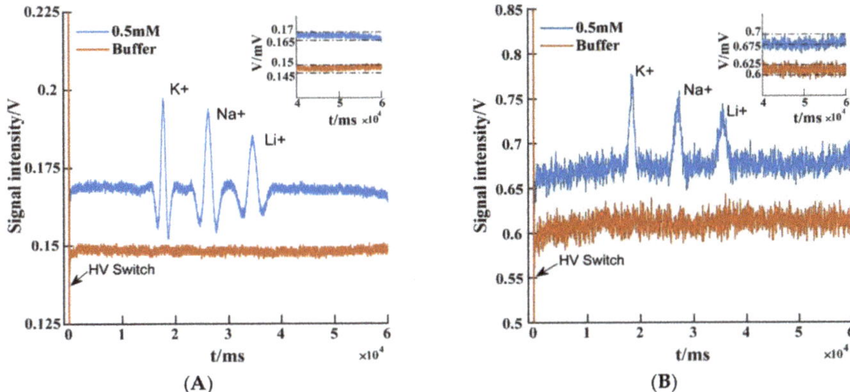

Figure 4. Raw electropherograms showing the separation of inorganic cations (K^+, Na^+, and Li^+, 0.5 mM each). (**A**) PG-C4D and (**B**) traditional C4D. Injection voltage: 500 V; separation voltage: 1000 V. Running buffer: 20 mM MES/His, pH 6.0. Detection conditions: 700-kHz, 5Vpp.

Table 2. Response Characteristics of PG-C4D and Traditional C4D (K^+, Na^+, and Li^+, 0.5 mM each).

Ion [1]	Indicator	PG-C4D (RSD)	Traditional C4D (RSD)
K^+	Migration Time (ms)	0.15%	0.24%
	Area (mV × ms)	1.56%	2.48%
	Height (mV)	0.43%	4.74%
Na^+	Migration Time (ms)	0.18%	0.39%
	Area (mV × ms)	1.57%	4.40%
	Height (mV)	1.94%	2.84%
Li^+	Migration Time (ms)	0.23%	0.63%
	Area (mV × ms)	2.88%	9.46%
	Height (mV)	3.42%	6.28%

[1] Data were filtered by Butterworth (cutoff frequency = 5 Hz) and obtained by trapezoidal numerical integration algorithm.

Compared with traditional C4D, the absolute peak height value of PG-C4D was reduced due to a slight loss of signal current. However, the signal-to-noise ratio was improved by 270%, 290%, and 270% for K^+, Na^+, and Li^+, respectively. For traditional C4D, a large part of the current measured by the electronics was actually introduced by stray capacitance. The stray capacitance current caused larger output signal of the C4D, but it actually enhanced the noise proportion, which was harmful to the sensitivity of the system.

It could be seen from Table 2 that the RSD of migration time of K^+, Na^+, and Li^+ increased slightly for both PG-C4D and traditional C4D, which was due to discrimination caused by electric injection [35,36]. The results in Table 2 demonstrated that the PG-C4D had better repeatability for quantitative analysis. For PG-C4D, the RSD for peak area was found to be 1.56% for K^+, 1.57% for Na^+, and 2.88% for Li^+, and the RSD for peak height was found to be 0.43% for K^+, 1.92% for Na^+, and 3.42% for Li^+. The results were on par with those obtained on referenced C4D with capillary electrophoresis [28] or complex top-bottom cell with precise shield of microchip electrophoresis [21].

3.3. Separation and Quantification of Cations

To demonstrate the quantitative performance of PG-C4D on cationic mixtures, five samples with different concentrations were tested. Figure 5A shows the raw electropherograms of the five concentration levels, under the optimal excitation signal conditions. The detection limits of K^+, Na^+, and Li^+ cations were 0.05 mM, 0.1 mM, and 0.1 mM, respectively. The limit of detection was defined as the concentration at which the ratio of the peak height to the baseline noise obtained by filtering ($S/N \geq 3$), as shown in Appendix C, Figure A2. The detection sensitivity here is relatively poor, compared with the C4D for microchip electrophoresis in a previous report [16]. However, under the same conditions, the detection limit of the proposed PG-C4D is much better than that of the traditional C4D. The detection limit of traditional C4D is around 0.25 mM, as shown in Appendix D, Figure A3. However, traditional C4D baseline noise fluctuates randomly and cannot be maintained at a stable level. The detection sensitivity of the proposed PG-C4D could be further improved by high voltage excitation signal [21,28] and a dedicated low-noise amplifier, but these devices would increase the cost of the total system [25,26].

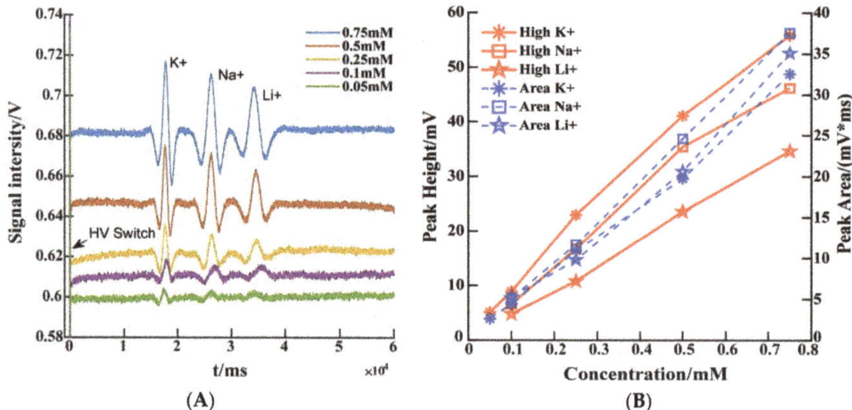

Figure 5. (**A**) An electropherogram of five concentrations. (**B**) Calibration curves of peak height sensitivity and peak area sensitivity of microchip electrophoresis with PG-C4D. Operational conditions were the same as in Figure 4.

The relation between the peak height/peak area and the concentration was investigated. The results showed that the peak height has higher concentration resolution. As shown in Figure 5B, the peak height/concentration correlation coefficients of K^+, Na^+, and Li^+ were $R^2 = 0.9931$, $R^2 = 0.9864$, and $R^2 = 0.9986$, respectively, and the corresponding sensitivities were 73.331 (K^+), 62.213 (Na^+), and 46.658 (Li^+) mV/mM, respectively. The correlation coefficients of the peak area/concentration of K^+, Na^+, and Li^+ were $R^2 = 0.995$, $R^2 = 0.9991$, and $R^2 = 0.9865$, respectively, and the corresponding sensitivities were 41.694 (K^+), 50.285 (Na^+), and 46.145 (Li^+) mV $\times$ ms/mM, respectively. These results are much better than those of the C4D with differential mode designed by Xiao [27] in capillary electrophoresis (such as, 2.2×10^{-3} (K^+), 2.3×10^{-3} (Li^+) mV/μM). The linear coefficients between the peak height against ion concentration and peak area against ion concentration of the same ion were almost the same. However, the sensitivity for peak height/concentration was higher than that of peak area/concentration. That is, peak height was a better indicator for concentration detection.

4. Conclusions

The reported PG-C4D had much lower stray capacitance than traditional C4D, thus the baseline intensity and noise amplitude of the detection cell were smaller, resulting in higher signal-to-noise ratio, detection sensitivity, and repeatability. Electrical field simulation and baseline strength test showed that the stray capacitance of PG-C4D was less than 1/3 of that of traditional C4D. The background baseline intensity of the PG-C4D was less than 1/6 of that of traditional C4D, and the noise amplitude of baseline was reduced from 25 mV to 4 mV. Electrophoretic detection experiments of inorganic cations showed that the RSD of migration time, peak area, and peak height were less than 0.3%, 3%, and 4%, respectively. The detection limits of inorganic K^+, Na^+, and Li^+ are 0.05, 0.1, and 0.1 mM, respectively. Experimental results showed that the designed PG-C4D can be used in a wide range of excitation frequencies and amplitudes, which was especially favorable for systems with large signal excitation.

This paper proposes a new C4D device for microchip electrophoresis with low stray capacitance, high sensitivity, and high repeatability, which requires no special shielding and chip size and shape requirement. At the same time, the microchip and electrodes are produced independently, and the microchip is easily replaced through mechanical assembly. PCB, CNC machining, and hot-press packaging greatly reduce development costs, which are of great significance in device integration and large-scale applications. The designed PG-C4D shows great potential for microchip electrophoresis, including the detection of industrial wastewater, agricultural irrigation water nutrients, and environmental conditions.

Author Contributions: Writing—original draft preparation, J.W.; planning and performing the experiments, J.W. and Y.L.; software support, W.H. and J.W.; theory consultation and formal analysis, W.H.; writing—review and editing, Y.C.; supervision, Y.C.; project administration, H.Y.; funding acquisition, H.Y. All authors have read and agreed to the published version of the manuscript.

Funding: This research was funded by Key-Area Research and Development Program of Guangdong Province (2019B020219003) and Guangxi Bagui Scholars Project (No.C3010099204).

Informed Consent Statement: Not applicable.

Data Availability Statement: Data is contained within the article.

Acknowledgments: Thanks to the teachers from the School of Mechanical Engineering of Guangxi University for their support in the research, as well as their opinions and suggestions on system construction and writing.

Conflicts of Interest: The authors declare no conflict of interest.

Appendix A

A model of the designed device geometry is shown in Figure A1. This device contains a 4-layer structure, the layers of which are microchannels, PMMA insulating layer, electrodes, and the PCB substrate, respectively.

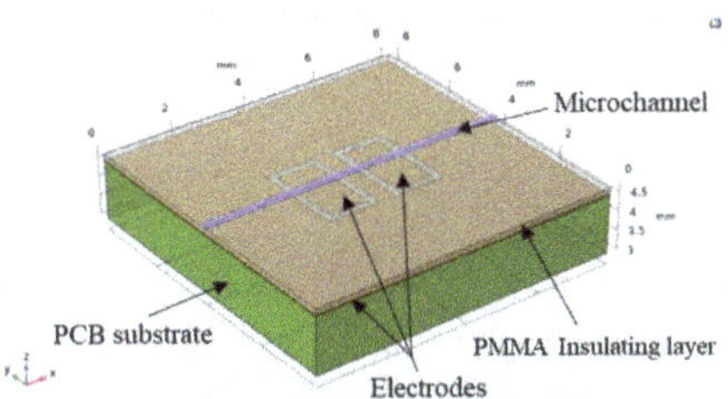

Figure A1. Three-dimensional model of the prototyped PG-C4D.

Appendix B

To calculate the capacitance, it was assumed that the sensing electrodes and the ground electrode form a mutual capacitance matrix. The capacitor matrix was composed of three capacitors, C_{g1}, C_{g2}, and C_L, as shown in Figure 3. The calculation principle was slightly more complicated: here is a brief overview. Potentials were applied to the two sensing electrodes, respectively, and the corresponding charges on the electrodes were measured. The next step was to get the mutual capacitance matrix by converting the Maxwell capacitance matrix. COMSOL used stationary source sweep/or manual terminal sweep to set one terminal to 1 V and all other terminals to ground.

In the drop-down list of calculation results, capacitance matrix with different symbols could be obtained through global matrix calculation. In this way, we could obtain the mutual capacitance matrix between the three electrodes, and then obtained the stray capacitance coupled between the sensing electrodes.

Appendix C

Figure A2 showed the filtered peaks and the fluctuation of baseline noise within 10 s. The filtered baseline noise was lower than 1.5 mV. It could be concluded that the filtering process can further improve the signal-to-noise ratio and it is beneficial for improving the detection limit of the system.

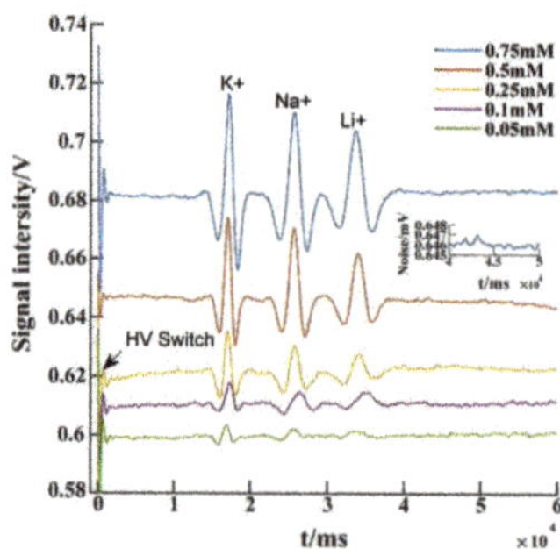

Figure A2. Signals filtered by Butterworth (cutoff frequency = 5 Hz), the inset diagram is the baseline noise fluctuation.

Appendix D

Three samples of different concentrations were tested on the traditional C4D to compare the performance of the proposed PG-C4D with the traditional C4D, as shown in Figure A3. When the sample concentration is 0.5 mM, the peak heights for K^+, Na^+, and Li^+ are about 94.5 mV, 75.7 mV, and 54.7 mV, respectively. When the sample concentration is 0.25 mM, the peak height for K^+ is 52.6 mV, and Na^+ 23.5 mV, while the peak height of Li^+ cannot be identified due to excessive signal noise. When the sample concentration is 0.1 mM, none of the three ionic ions can be identified. These results also showed that the fluctuation amplitude of the baseline noise is unstable, which indicates that the traditional C4D had strong instability.

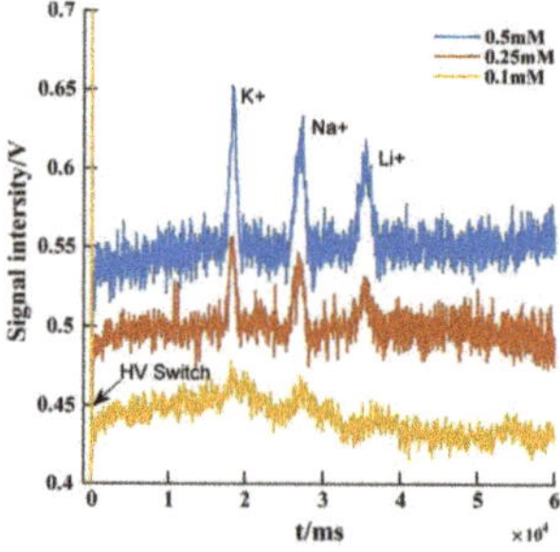

Figure A3. An electropherogram of three concentrations of device with traditional C4D. Operational conditions were the same as in Figure 5.

References

1. Hauser, P.C.; Kuban, P. Capacitively coupled contactless conductivity detection for analytical techniques—Developments from 2018 to 2020. *J. Chromatogr. A* **2020**, *1632*, 461616. [CrossRef]
2. Zhang, M.; Phung, S.C.; Smejkal, P.; Guijt, R.M.; Breadmore, M.C. Recent trends in capillary and micro-chip electrophoretic instrumentation for field-analysis. *Trends Environ. Anal. Chem.* **2018**, *18*, 1–10. [CrossRef]
3. Ou, X.; Chen, P.; Huang, X.; Li, S.; Liu, B.-F. Microfluidic chip electrophoresis for biochemical analysis. *J. Sep. Sci.* **2020**, *43*, 258–270. [CrossRef]
4. Le, T.B.; Hauser, P.C.; Pham, T.N.M.; Kieu, T.L.P.; Le, T.P.Q.; Hoang, Q.A.; Le, D.C.; Nguyen, T.A.H.; Mai, T.D. Low-cost and versatile analytical tool with purpose-made capillary electrophoresis coupled to contactless conductivity detection: Application to antibiotics quality control in Vietnam. *Electrophoresis* **2020**, *41*, 1980–1990. [CrossRef] [PubMed]
5. Zemann, A.J.; Mayrhofer, K.; Schnell, E.; Bonn, G.K. Contactless conductivity detection for capillary electrophoresis. *Abstr. Pap. Am. Chem. Soc.* **1998**, *216*, U162. [CrossRef]
6. Da Silva, J.A.F.; do Lago, C.L. An oscillometric detector for capillary electrophoresis. *Anal. Chem.* **1998**, *70*, 4339–4343. [CrossRef]
7. Pumera, M.; Wang, J.; Opekar, F.; Jelinek, I.; Feldman, J.; Lowe, H.; Hardt, S. Contactless conductivity detector for microchip capillary electrophoresis. *Anal. Chem.* **2002**, *74*, 1968–1971. [CrossRef] [PubMed]
8. Gamat, S.N.; Fotouhi, L.; Talebpour, Z. The application of electrochemical detection in capillary electrophoresis. *J. Iran. Chem. Soc.* **2017**, *14*, 717–725. [CrossRef]
9. Brito-Neto, J.G.A.; da Silva, J.A.F.; Blanes, L.; do Lago, C.L. Understanding capacitively coupled contactless conductivity detection in capillary and microchip electrophoresis. Part 1. Fundamentals. *Electroanalysis* **2005**, *17*, 1198–1206. [CrossRef]
10. Brito-Neto, J.G.A.; da Silva, J.A.F.; Blanes, L.; do Lago, C.L. Understanding capacitively coupled contactless conductivity detection in capillary and microchip electrophoresis. Part 2. Peak shape, stray capacitance, noise, and actual electronics. *Electroanalysis* **2005**, *17*, 1207–1214. [CrossRef]
11. Kuban, P.; Hauser, P.C. Ten years of axial capacitively coupled contactless conductivity detection for CZE—A review. *Electrophoresis* **2009**, *30*, 176–188. [CrossRef] [PubMed]
12. Kuban, P.; Hauser, P.C. Fundamental aspects of contactless conductivity detection for capillary electrophoresis. Part I: Frequency behavior and cell geometry. *Electrophoresis* **2004**, *25*, 3387–3397. [CrossRef] [PubMed]
13. Koczka, P.I.; Bodoki, E.; Gaspar, A. Application of capacitively coupled contactless conductivity as an external detector for zone electrophoresis in poly(dimethylsiloxane) chips. *Electrophoresis* **2016**, *37*, 398–405. [CrossRef] [PubMed]
14. Takekawa, V.S.; Marques, L.A.; Strubinger, E.; Segato, T.P.; Bogusz, S., Jr.; Brazaca, L.C.; Carrilho, E. Development of low-cost planar electrodes and microfluidic channels for applications in capacitively coupled contactless conductivity detection ((CD)-D-4). *Electrophoresis* **2021**, *42*, 1560–1569. [CrossRef] [PubMed]
15. Liu, J.; Xu, F.; Wang, S.; Chen, Z.; Pan, J.; Ma, X.; Jia, X.; Xu, Z.; Liu, C.; Wang, L. A polydimethylsiloxane electrophoresis microchip with a thickness controllable insulating layer for capacitatively coupled contactless conductivity detection. *Electrochem. Commun.* **2012**, *25*, 147–150. [CrossRef]
16. Kuban, P.; Hauser, P.C. Evaluation of microchip capillary electrophoresis with external contactless conductivity detection for the determination of major inorganic ions and lithium in serum and urine samples. *Lab Chip* **2008**, *8*, 1829–1836. [CrossRef]
17. Guijt, R.M.; Evenhuis, C.J.; Macka, M.; Haddad, P.R. Conductivity detection for conventional and miniaturised capillary electrophoresis systems. *Electrophoresis* **2004**, *25*, 4032–4057. [CrossRef]
18. Wang, Y.; Cao, X.; Hogan, A.; Messina, W.; Moore, E.J. Fabrication of a grounded electrodes cell with Capacitively Coupled Contactless Conductivity Detection technique in a microchip capillary electrophoresis application. In Proceedings of the 18th IEEE International Conference on Nanotechnology (IEEE-NANO), Tyndall Natl Inst, Cork, Ireland, 23–26 July 2018.
19. Petkovic, K. An integrated portable multiplex microchip device for fingerprinting chemical warfare agents. Micromachines. *Micromachines* **2019**, *10*, 617. [CrossRef]
20. Guijt, R.M.; Armstrong, J.P.; Candish, E.; Lefleur, V.; Percey, W.J.; Shabala, S.; Hauser, P.C.; Breadmore, M.C. Microfluidic chips for capillary electrophoresis wionth integrated electrodes for capacitively coupled conductivity detection based on printed circuit board technology. *Sens. Actuators B Chem.* **2011**, *159*, 307–313. [CrossRef]
21. Mahabadi, K.A.; Rodriguez, I.; Lim, C.Y.; Maurya, D.K.; Hauser, P.C.; de Rooij, N.F. Capacitively coupled contactless conductivity detection with dual top-bottom cell configuration for microchip electrophoresis. *Electrophoresis* **2010**, *31*, 1063–1070. [CrossRef]
22. Kang, Q.; Shen, D.; Li, Q.; Hu, Q.; Dong, J.; Du, J.; Tang, S. Reduction of the impedance of a contactless conductivity detector for microchip capillary electrophoresis: Compensation of the electrode impedance by addition of a series inductance from a piezoelectric quartz crystal. *Anal. Chem.* **2008**, *80*, 7826–7832. [CrossRef]
23. Zhang, Z.; Li, D.; Liu, X.; Subhani, Q.; Zhu, Y.; Kang, Q.; Shen, D. Determination of anions using monolithic capillary column ion chromatography with end-to-end differential contactless conductometric detectors under resonance approach. *Analyst* **2012**, *137*, 2876–2883. [CrossRef]
24. Huang, J.; Ji, H.; Huang, Z.; Wang, B.; Li, H. A New Contactless Method for Velocity Measurement of Bubbleand Slug in Millimeter-Scale Pipelines. *IEEE Access* **2017**, *5*, 12168–12175. [CrossRef]
25. Laugere, F.; Guijt, R.M.; Bastemeijer, J.; van der Steen, G.; Berthold, A.; Baltussen, E.; Bossche, A. On-chip contactless four-electrode conductivity detection for capillary electrophoresis devices. *Anal. Chem.* **2003**, *75*, 306–312. [CrossRef] [PubMed]

26. Fercher, G.; Haller, A.; Smetana, W.; Vellekoopt, M.J. End-to-End Differential Contact less Conductivity Sensor for Microchip Capillary Electrophoresis. *Anal. Chem.* **2010**, *82*, 3270–3275. [CrossRef]
27. Wang, C.; Xing, H.; Zheng, B.; Yuan, H.; Xiao, D. Simulation and Experimental Study on Doubled-Input Capacitively Coupled Contactless Conductivity Detection of Capillary Electrophoresis. *Sci. Rep.* **2020**, *10*, 7944. [CrossRef] [PubMed]
28. Stojkovic, M.; Schlensky, B.; Hauser, P.C. Referenced Capacitively Coupled Conductivity Detector for Capillary Electrophoresis. *Electroanalysis* **2013**, *25*, 2645–2650. [CrossRef]
29. Chang, Y.; You, H. A hybrid adhesive bonding of PMMA and PCB with an application on microchip electrophoresis. *Anal. Methods* **2019**, *11*, 1229–1236. [CrossRef]
30. Liu, B.; Jin, Q.; Zhang, Y.; Mayer, D.; Krause, H.-J.; Zhao, J.; Offenhaeusser, A. A simple poly(dimethylsiloxane) electrophoresis microchip with an integrated contactless conductivity detector. *Microchim. Acta* **2011**, *172*, 193–198. [CrossRef]
31. Liu, B.; Zhang, Y.; Mayer, D.; Krause, H.-J.; Jin, Q.; Zhao, J.; Offenhäusser, A.; Xu, Y. Determination of heavy metal ions by microchip capillary electrophoresis coupled with contactless conductivity detection. *Electrophoresis* **2012**, *33*, 1247–1250. [CrossRef] [PubMed]
32. Huang, Z.; Yang, M.; You, H.; Xie, Y. Simultaneous Determination of Inorganic Cations and Anions in Microchip Electrophoresis Using High-voltage Relays. *Anal. Sci.* **2018**, *34*, 801–805. [CrossRef] [PubMed]
33. How to Calculate the Capacitance Matrix in COMSOL Multiphysics®. Available online: http://cn.comsol.com/blogs/how-to-calculate-a-capacitance-matrix-in-comsol-multiphysics (accessed on 17 June 2017).
34. Lichtenberg, J.; de Rooij, N.F.; Aeroporto, E. A microchip electrophoresis system with integrated in-plane electrodes for contactless conductivity detection. *Electrophoresis* **2002**, *23*, 3769–3780. [CrossRef]
35. Gaudry, A.J.; Nai, Y.H.; Guij, R.M.; Breadmore, M.C. Polymeric Microchip for the Simultaneous Determination of Anions and Cations by Hydrodynamic Injection Using a Dual-Channel Sequential Injection Microchip Electrophoresis System. *Anal. Chem.* **2014**, *86*, 3380–3388. [CrossRef] [PubMed]
36. Ha, N.S.; Ly, J.; Jones, J.; Cheung, S.; van Dam, R.M. Novel volumetric method for highly repeatable injection in microchip electrophoresis. *Anal. Chim. Acta* **2017**, *985*, 129–140. [CrossRef] [PubMed]

 micromachines

Article

Microfluidics Integration into Low-Noise Multi-Electrode Arrays

Mafalda Ribeiro [1,2], Pamela Ali [2], Benjamin Metcalfe [2], Despina Moschou [2,*] and Paulo R. F. Rocha [3,*]

1 Centre for Accountable, Responsible, and Transparent AI (ART-AI), Department of Computer Science, University of Bath, Bath BA2 7AY, UK; mr611@bath.ac.uk

2 Centre for Biosensors, Bioelectronics, and Biodevices (C3Bio), Department of Electronic and Electrical Engineering, University of Bath, Bath BA2 7AY, UK; pamela7ali@gmail.com (P.A.); B.W.Metcalfe@bath.ac.uk (B.M.)

3 Centre for Functional Ecology (CFE), Department of Life Sciences, University of Coimbra, 3000-456 Coimbra, Portugal

* Correspondence: D.Moschou@bath.ac.uk (D.M.); procha@uc.pt (P.R.F.R.); Tel.: +44-(0)-1225-383245 (D.M.); +351-239-240-700 (P.R.F.R.)

Abstract: Organ-on-Chip technology is commonly used as a tool to replace animal testing in drug development. Cells or tissues are cultured on a microchip to replicate organ-level functions, where measurements of the electrical activity can be taken to understand how the cell populations react to different drugs. Microfluidic structures are integrated in these devices to replicate more closely an in vivo microenvironment. Research has provided proof of principle that more accurate replications of the microenvironment result in better micro-physiological behaviour, which in turn results in a higher predictive power. This work shows a transition from a no-flow (static) multi-electrode array (MEA) to a continuous-flow (dynamic) MEA, assuring a continuous and homogeneous transfer of an electrolyte solution across the measurement chamber. The process through which the microfluidic system was designed, simulated, and fabricated is described, and electrical characterisation of the whole structure under static solution and a continuous flow rate of 80 μL/min was performed. The latter reveals minimal background disturbance, with a background noise below 30 μVpp for all flow rates and areas. This microfluidic MEA, therefore, opens new avenues for more accurate and long-term recordings in Organ-on-Chip systems.

Keywords: MEA; brain cells; electrical recordings; Organ-on-Chip; microfluidics; Brain-on-Chip

Citation: Ribeiro, M.; Ali, P.; Metcalfe, B.; Moschou, D.; Rocha, P.R.F. Microfluidics Integration into Low-Noise Multi-Electrode Arrays. *Micromachines* **2021**, *12*, 727. https://doi.org/10.3390/mi12060727

Academic Editor: Francisco Perdigones

Received: 5 May 2021
Accepted: 17 June 2021
Published: 20 June 2021

Publisher's Note: MDPI stays neutral with regard to jurisdictional claims in published maps and institutional affiliations.

1. Introduction

The high cost of drug development commonly arises from the low success rate of clinically applicable drugs during the development stage. It takes on average 10–12 years [1] to develop a new drug, with two-thirds of the total costs ascribed to the clinical trial stage [2]. In order to reduce these costs and, thus, allow the more rapid introduction of effective drugs in clinical practice, it is critical that the accuracy and throughput of pre-clinical screening is significantly improved [3]. The traditional drug development route mostly relies on animal testing, which is costly, ethically contentious, and often offers poor predictive power for human response to drugs due to interspecific discrepancies [4].

Currently, the predictive power of alternative tissue models is compromised due to oversimplified cell microenvironments and tissue structures [5]. Micro-physiological systems, also known as Organ-on-Chip (OOC) technology, integrate microscale organ models on microchips, connected with microfluidic channels [6], delivering nutrients and reagents on-chip precisely and continuously, replicating tissue interactions within the body [7]. The in vitro cell culturing and microfluidics that are employed in this technology can emulate drug absorption, distribution, and metabolism in the human body much more accurately, increasing the potential of successful drug development. According to Wang et al. [5], the five core aspects of developing accurate OOC systems are a combination of the human physiology modelling and the microenvironment in which the cells are being cultured.

These five areas are the controlled culture environment, pumpless microfluidic platforms, functional measurements, advanced single-organ models, and system integration.

Owing to its game-changing potential, OOC technology is currently being explored for various organ models. Brain-on-Chip technology, in particular, attempts to model and measure high-level brain functions which emerge from the interaction of interconnected neural networks [8]. Haring et al. highlighted the limitations of reductionist approaches and how they often fail to replicate the complexity and higher-order features of neural networks and the human nervous system.

Measurements from the neural system are based on the electrical signalling associated with fluctuations of the membrane potential in individual cells [9]. These oscillations are a result of the bidirectional flow of Na^+ and K^+ ions in neurons, resulting in membrane potentials which reach tens of mVs when recorded intracellularly or hundreds of μVs when recorded extracellularly. Multi-electrode arrays (MEA) are used to measure the extracellular local field potentials of the cells that are adhered to microchip electrodes [10]. Using MEAs to measure extracellular bioelectrical activity in vitro enables the study of neuronal network processes, the electrophysiological mechanisms related to pathological diseases, and the effects of drugs on the cell populations [11]. Nonetheless, although neuronal firing is routinely recorded, extremely low-magnitude electrical signals, below 50 μV, still pose significant challenges in MEA recordings.

An ultra-sensitive MEA capable of detecting the electrical activity of electrically quiescent cells such as rat C6 Glioma cells [9], breast cancer [12], and prostate cancer [13] has been previously developed. The microchip was made of a glass substrate with gold (Au) circular electrodes with a polymethyl methacrylate (PMMA) well glued on top of a substrate, serving as a container for the electrolyte solution. In addition to glass-based substrates, printed circuit board (PCB) approaches have also been implemented in the past to improve cost effectiveness, scalability to mass production, and future electronics integration of electrochemical devices [14]. Despite these advantages, it is important to note that PCBs traditionally use copper (Cu) for creating conductive traces and electrodes. A surface coating of Au is then deposited, as a standard process for board viability [15]. Although this improves biocompatibility, previous studies have indicated that exposed Cu-based structures typically show an increase in electrochemical noise due to corrosion, in relation to homogeneous Au surfaces [16,17]. The impact of electrochemical noise in standard PCB-based electrodes, in the context of large-area MEAs, has not been previously investigated.

This work demonstrates for the first time a transformation of a static, large-electrode-area MEA to a continuous flow, low-noise system. A custom-made microfluidic network was designed, enabling efficient nutrient delivery and waste removal across the whole electrode area. Prototype fabrication and fluidic functionality of the device are outlined, exploiting a PCB version of the previously demonstrated MEA aiming to leverage the associated low-cost production and integration. Section 2 shows the methodology behind the design and manufacturing process of the microfluidic MEA, as well as the proposed setup for conducting noise recordings. Section 3 highlights the implementation of the final prototype and initial electrical characterisation of the device. Sections 4 and 5 contain a discussion of the results, including future work required.

2. Materials and Methods

Different designs for a microfluidic network compatible with the geometrical characteristics of a large-area MEA and electrical measurement setup were generated using commercial CAD software (AutoCAD® 22.0). The software package COMSOL Multiphysics® 5.3a was used to simulate the concentration profiles of the designed microfluidic structures under steady flow and select the one resulting in the largest uniformity. For all models, COMSOL stationary studies were used for creeping flow and transport of diluted species. The inlet flow rate, concentration, diffusion constant, and viscosity of the liquid were selected to represent experimental conditions for the flow of the growth medium for cells

(100 μL/min, 1 mM, 1×10^{-9} m^2/s, and 0.5 m^3/mol^2, respectively). An extra-coarse mesh was exploited with a "Free Quad" symmetry face, optimised for fluid dynamic studies (maximum element size 600, minimum element size 120, maximum element growth rate 1.3, curvature factor 0.9, and resolution of narrow regions 0.4).

Once the optimum microfluidic design was selected, the structure design was rapid prototyped and tested for fluidic tightness. A 0.05 mm thick PMMA sheet was sandwiched between two layers of double-sided adhesive film (3M 468MP). The final microfluidic design was laser-cut out onto the 0.5 mm PMMA and adhesive assembly. A lid, with the corresponding input and output hole, was laser-cut on 5 mm thick PMMA. The patterned PMMA and adhesive structure were stuck onto the 5 mm lid by removing one side of the double-sided adhesive film. This assembly was then adhered onto the MEA micro-chip by removing the second side of the film. The MEA device was designed in a standard PCB design CAD software (Altium Designer ® 17.0), comprising a gold-plated sensing electrode layer. The sensing layer consisted of eight planar, circular electrodes of four different surface areas. The dimensions of the MEA were replicated from a previously exploited MEA on glass substrate, so that the same electrical interfacing apparatus could be exploited for low-noise brain cell signal recordings [10]. A PCB-based substrate was used in this instance instead of glass due to the cost effectiveness, potential for large-scale manufacturing, and ease for further electronics integration. The surface roughness as measured with a DEKTAK surface profilometer amounted to approximately 100 nm. Hence, the electrode area could be taken equal to the geometrical value of the circular electrodes. The areas used amounted to 1 mm^2, 2 mm^2, 7 mm^2, and 12 mm^2. The glass-reinforced epoxy laminate FR4 had a thickness of 800 μm, and it supported an Au plated layer above 35 μm of Cu. The functionality of the completed prototype was tested using a dye solution injected through the device inlet to verify its fluidic tightness, as well as uniform coverage of all electrodes.

Electrical voltage noise recordings were carried out on the manufactured chips using static and continuously flowing KCl solution. The conductivity of 100 mM aqueous KCl was calculated in a previous study to be approximately 200 Ω·cm, which is in close agreement with Sigma Aldrich's reference conductivity value for culture medium [10]. The solution resistance, modelled in detail previously, decreases when the molarity increases, which leads to a gradual change of the Maxwell–Wagner relaxation frequency for OOC systems. Hence, 100 mM KCl solution was also used in this work as an approximation to cell culture medium for the purpose of electrical noise measurements. Flow rate control was achieved with a microfluidics system, consisting of a pressure pump (OB1 MK3+; Elveflow, Paris, France [18]) and digital flow sensor (MFS3; Elveflow, Paris, France [19]) capable of measuring flow rates up to 80 μL/min. Similarly, prior microfluidics systems have used flow rates ranging from 1 μL to 1 mL/min [20–22]. The microfluidic MEA was placed in a container and connected to a voltage amplifier (Brookdeal Preamplifier 5006; Ortec, TN, USA), high-speed analogue-to-digital converter (ADC) (NI USB-6343; National Instruments, Austin, TX, USA), and a device for minimising mains (50 Hz) interference (Humbug; Quest Scientific, Vancouver, BC, Canada). The chip container, flow sensor, and voltage amplifier were contained within a Faraday cage to minimise interference in the recordings. Voltage recordings were conducted at a sampling rate of 125 kHz and an amplifier gain of 1075. The recordings were downsampled to 50 kHz for offline analysis and the power spectra were further filtered with a moving mean (window length of 5) for clarity.

3. Results

3.1. Concentration Profile

The microfluidic MEA developed for the purposes of this study was designed to ensure both continuous and uniform flow of growth medium across the complete surface of the chip. Alternative geometries were designed via COMSOL, and the concentration profile was modelled considering an inlet flow rate of 100 μL/min. The initial design comprised a single inlet and outlet, positioned in the middle of the opposite sides of the chamber (Figure 1a). This configuration resulted in accumulation of liquid at the corners and edges of the MEA area. This nonuniformity in the cell culture medium flow over the MEA surface is expected to be particularly detrimental for future cell-based studies. In particular, this design could result in cell death on the corners and edges of the MEA due to the effects of cell metabolism, depletion of nutrients at the corners of the device, and excess waste accumulation. To mitigate this issue, two additional inlets were added symmetrically to the original one (Figure 1b).

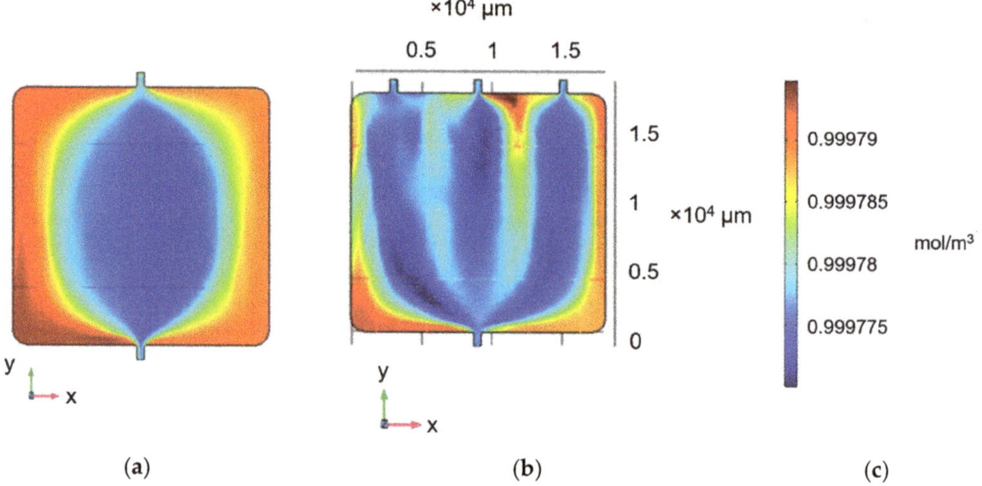

Figure 1. Concentration profile (mol/m^3) of square continuous-flow chamber featuring (**a**) one inlet and one outlet, and (**b**) three inlets and one outlet at steady state. (**c**) Concentration range for (**a**,**b**).

Although the accumulation of liquid at the top two corners was reduced, there were still low-flow spots between the inlets and at the bottom two corners. Considering the addition of more inlets impractical for implementation, the three-inlet approach was chosen, but the chamber shape was modified from square to a trapezium, following the MEA electrode pattern and theoretically minimising accumulation in the bottom corners. Two alternative versions were simulated, one following, in direct proportionality, the electrode contour (Figure 2a) and one with straight sides tangentially covering their complete surface (Figure 2b).

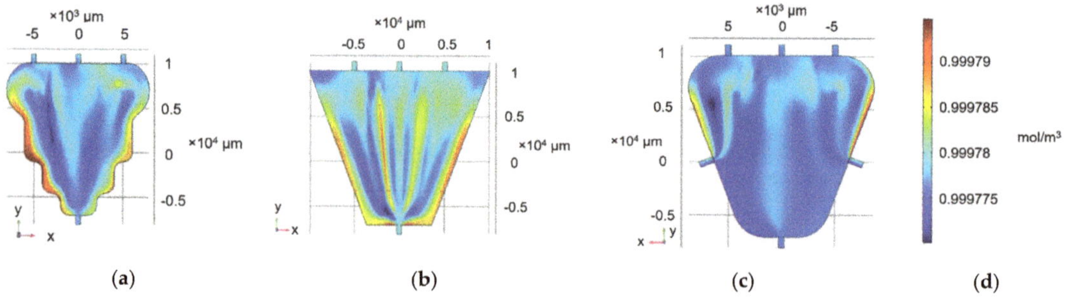

Figure 2. Concentration profile (mol/m^3) of trapezium continuous-flow chamber featuring three inlets and one outlet (**a**) with a curved profile following electrode geometry and (**b**) straight edges (**c**) featuring three inlets and three outlets. (**d**) Concentration range for (**a**,**b**,**c**).

In Figure 2a,b, the accumulation of liquid in the bottom was minimised; nonetheless, there was slight accumulation in the edge of the trapezium side faces. For this reason, a final modification was simulated (Figure 2c). The three-inlet trapezium approach was maintained; however, rounding the edges with a 3 mm radius facilitated a smoother flow, and adding two symmetrical outlets in the middle of the side edges minimised liquid accumulation on the edges.

3.2. Microfluidic Prototype Implementation

The identified optimal microfluidic chamber geometry was implemented in the final device prototype. The final device should also be compatible with the low-noise laboratory measurement setup. To this end, (1) the height of the device could not exceed 30 mm, (2) there had to be an inlet and outlet hole within chip area, and (3) the inlet and outlet channels had to be of equal length and width, to assure synchronised liquid flow over the chip.

A model of the chosen device geometry and materials are shown in Figure 3. For this device, the design depicted in Figure 4a was developed, housing two standard Luer-type microfluidic connectors for the efficient interfacing of the chip to the macroscopic world. This design was laser-micromachined in the PMMA double-sided tape sandwich described in Section 2. The micromachined pattern was then adhered on the PCB MEA substrate and finally sealed with the 5 mm PMMA lid, housing the inlets for the microfluidic connectors. The connectors were screwed in the inlets and tubings attached to them to deliver the dye solution. On some instances, there was accumulation of bubbles in several locations on the chip. When this was observed, the device was placed in an O_2 plasma chamber (Zepto Model 2, Diener Electronic, Ebhausen, Germany) and treated for 15 s at 100 mW in order to become more hydrophilic and facilitate bubble-free initial wetting of the chip surface (Figure 4b–d).

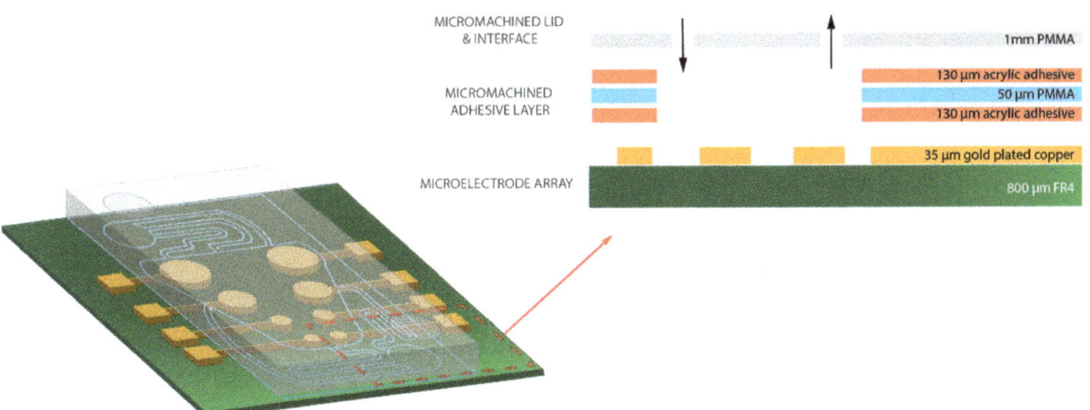

Figure 3. Three-dimensional model of the prototyped continuous flow microfluidic MEA platform. The sensing part of the device consists of four electrode pairs with the areas 12, 7, 2, and 1 mm^2, and electrode spacing of 9, 7, 3, and 2 mm, respectively (measured from the centres of each electrode). The inset shows a magnified view of the highlighted cross-section, in dashed red lines, comprising the following material layers: FR4 (green), gold-plated contacts and electrodes (yellow), acrylic adhesive (orange), 50 µm PMMA (blue), and 1 mm PMMA for the lid (gray).

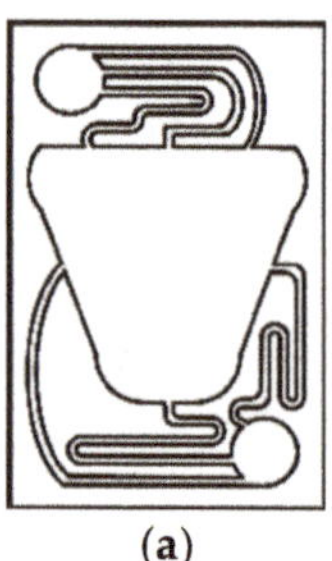
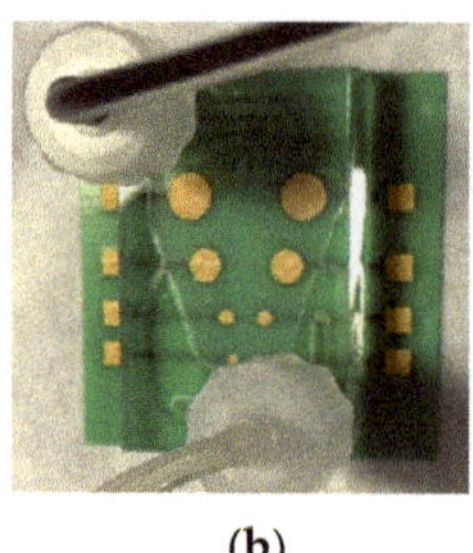
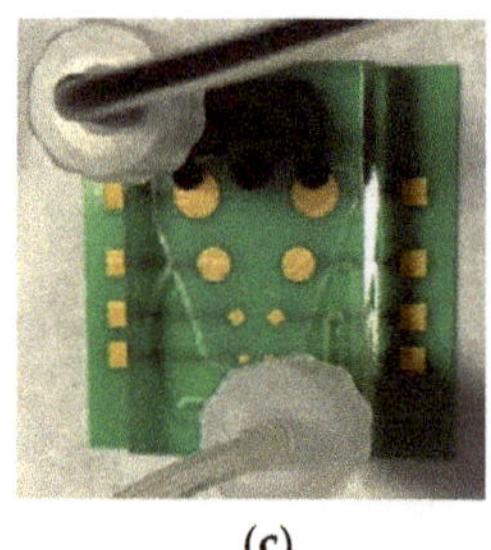
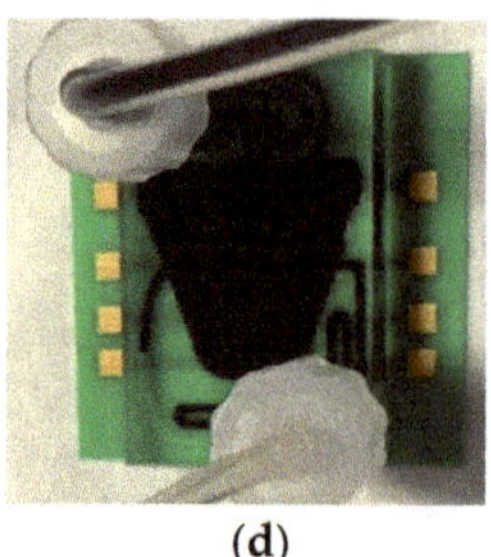

(a) (b) (c) (d)

Figure 4. (**a**) Microfluidic network design. (**b–d**) Assembled microfluidic MEA tested for fluidic tightness under continuous flow using a dye (**b**) when the device chamber was empty, (**c**) when the device chamber filled through the inlet ports, and (**d**) when the dye moved through the centre and into the outlet ports.

3.3. Electrical Characterisation

Voltage recordings were conducted on all electrode areas at zero flow (0 μL/min) and the maximum sensor flow rate (80 μL/min) using the setup shown in Figure 5a. No significant change in electrical voltage noise was seen when transitioning from no flow to flow (this change varied between 1.2×10^{-10} and 1.9×10^{-9} V^2/Hz). Power spectra for each area at a flow rate of 80 μL/min are shown in Figure 5b. At thermodynamic equilibrium, at zero volt overpotential, the voltage power spectral density is given as follows [10]:

$$S_V(\omega) = 4kTZ'(\omega) \tag{1}$$

where Z' is the real component of the impedance, T is the temperature, k is the Boltzmann constant, and ω is the angular frequency. Given that the electrodes in question are not ideally polarizable, the double-layer capacitance was described using a constant phase element (CPE), and the impedance as a function of CPE was then used to model the expected noise of an electrode/electrolyte system. This was investigated for the case of a metal/electrolyte interface for Au/Electrolyte in prior work [10]. According to Equation (1), with increasing electrode area and, hence, decreasing impedance, voltage power spectral density should decrease. This is highlighted in the inset of Figure 5b with the quantified voltage power at 20 Hz. These results indicate that the Cu/Au electrodes showed a larger baseline noise level in comparison to Cr/Au electrodes on a glass substrate; however, despite this, the voltage noise was kept, at all times, below 30 μVpp (with amplifier RMS noise of 1.13 μVpp at a bandwidth of 125 kHz). Although mains interference was minimised using the Humbug and Faraday cage, it is possible that other interfering signals may have been present in the recordings from surrounding equipment, given that the Faraday cage could not be completely shut with microfluidic tubing and electrical wiring in place, as well as that the ADC and Humbug had to be placed outside the cage.

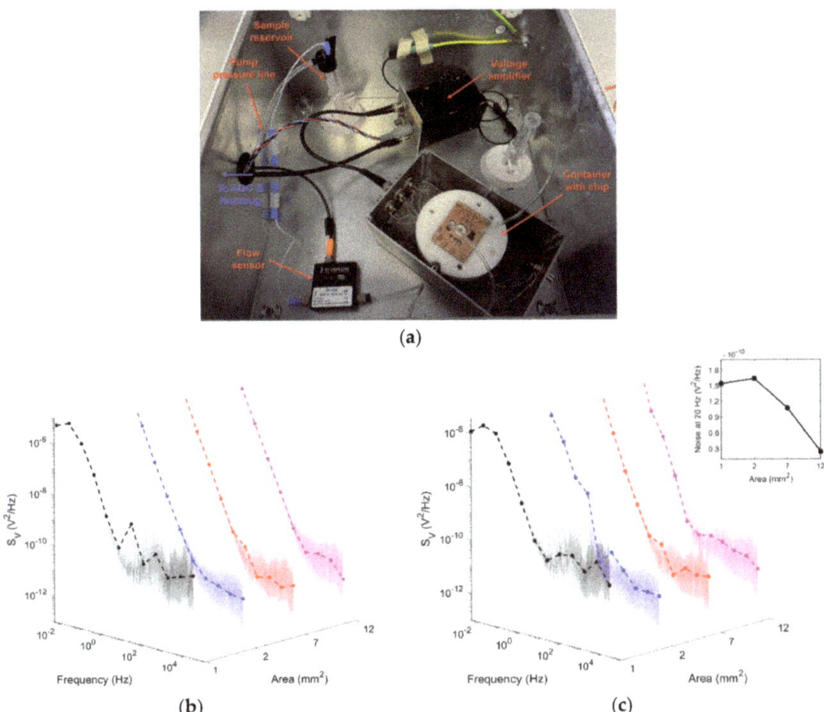

Figure 5. (**a**) Microfluidics setup within Faraday cage, which includes the sample reservoir, microfluidic flow sensor (MFS3; Elveflow), voltage amplifier (Brookdeal Preamplifier 5006; Ortec), and a custom-built container for the MEA and microfluidic tubing. The chip container was covered with aluminium foil for experiments and the Faraday cage was closed. (**b**,**c**) Power spectra of all electrode areas (**b**) under a static solution, and (**c**) at a continuous flow rate of 80 μL/min. Inset: Change in noise at 20 Hz across all electrode areas under continuous flow.

4. Discussion

In order to transition from static MEA structures to continuous-flow ones, the design of the microfluidic structure needs to be carefully selected, assuring a uniform flow across the complete chip surface; nonuniformities in flow could result in preferential cell growth in specific parts of the array [23], thus biasing the electrical recordings. The practical aspects need to also be carefully considered, enabling the exploitation of low-noise measurement setups for reliable cell signal recordings. It is also very critical to assure a bubble-free operation under flow since the presence of bubbles over the electrodes will compromise the signal transduction. In this work, a systematic study was described which efficiently achieves all of the above, exploiting a PCB-based MEA. Electrical characterisation tests were performed on the chip under both static and continuous flow, indicating that an increase in flow rate did not significantly influence the recoding background noise, where a plateau of 30 μVpp was maintained. Future work includes modelling nutrient consumption within the system to ascertain how parameters such as velocity, wall shear stress, and critical perfusion rate are affected and, hence, the suitability of the device to different cell types [24,25]. These simulations should be run in tandem with experimental cell work, given the inherent complexity associated with modelling cells. Lastly, biocompatibility studies should be performed using cellular cultures, and further electrical characterisation over a broader range of flow rates should be conducted.

5. Conclusions

In this work, the transition from static MEAs to continuous-flow ones, towards more realistic Brain-on-Chip structures, was described. Different microfluidic designs were simulated, aiming for uniform growth medium distribution across a previously demonstrated, custom-made MEA for cell electrical sensing. The optimum design comprised a three-inlet, three-outlet rounded-corner microfluidic chamber following the electrode periphery. The design was integrated on a PCB-fabricated MEA for the first time, and a plateau below 30 μVpp was achieved when changing from a static solution to a maximum flow rate of 80 μL/min. The observed change in voltage noise was minimal when changing from a static KCl solution to a continuously flowing one. Future work will include electrical characterisation of the device under continuous flow in a wider range of flow rates along with signal recordings with electrogenic cells.

Author Contributions: Conceptualisation, P.R.F.R. and D.M.; methodology, D.M., P.R.F.R., B.M., and P.A.; software, M.R.; validation, M.R.; resources, P.R.F.R., B.M., and D.M.; data curation, M.R.; writing—original draft preparation, P.R.F.R., D.M., and B.M.; writing—review and editing, P.A., M.R., and B.M. All authors read and agreed to the published version of the manuscript.

Funding: This research was funded by The Royal Society, grant number RSG\R1\180260 "Modelling brain function through Organ-on-Chip platforms for drug discovery applications".

Conflicts of Interest: The authors declare no conflict of interest.

References

1. DiMasi, J.A.; Grabowski, H.G.; Hansen, R.W. Innovation in the pharmaceutical industry: New estimates of R&D costs. *J. Health Econ.* **2016**, *47*, 20–33. [PubMed]
2. Paul, S.M.; Mytelka, D.S.; Dunwiddie, C.T.; Persinger, C.C.; Munos, B.H.; Lindborg, S.R.; Schacht, A.L. How to improve RD productivity: The pharmaceutical industry's grand challenge. *Nat. Rev. Drug Discov.* **2010**, *9*, 203–214. [CrossRef] [PubMed]
3. Zhang, B.; Radisic, M. Organ-on-A-chip devices advance to market. *Lab Chip* **2017**, *17*, 2395–2420. [CrossRef]
4. Eastwood, D.; Findlay, L.; Poole, S.; Bird, C.; Wadhwa, M.; Moore, M.; Burns, C.; Thorpe, R.; Stebbings, R. Monoclonal antibody TGN1412 trial failure explained by species differences in CD28 expression on CD4 + effector memory T-cells. *Br. J. Pharmacol.* **2010**, *161*, 512–526. [CrossRef]
5. Wang, Y.I.; Oleaga, C.; Long, C.J.; Esch, M.B.; McAleer, C.W.; Miller, P.G.; Hickman, J.J.; Shuler, M.L. Self-contained, low-cost Body-on-a-Chip systems for drug development. *Exp. Biol. Med.* **2017**, *242*, 1701–1713. [CrossRef] [PubMed]
6. Esch, E.W.; Bahinski, A.; Huh, D. Organs-on-chips at the frontiers of drug discovery. *Nat. Rev. Drug Discov.* **2015**, *14*, 248–260. [CrossRef]
7. Esch, M.B.; King, T.L.; Shuler, M.L. The role of body-on-a-chip devices in drug and toxicity studies. *Annu. Rev. Biomed. Eng.* **2011**, *13*, 55–72. [CrossRef]
8. Habibey, R.; Latifi, S.; Mousavi, H.; Pesce, M.; Arab-Tehrany, E.; Blau, A. A multielectrode array microchannel platform reveals both transient and slow changes in axonal conduction velocity. *Sci. Rep.* **2017**, *7*, 1–14. [CrossRef]
9. Rocha, P.R.F.; Schlett, P.; Schneider, L.; Dröge, M.; Mailänder, V.; Gomes, H.L.; Blom, P.W.M.; De Leeuw, D.M. Low frequency electric current noise in glioma cell populations. *J. Mater. Chem. B* **2015**, *3*, 5035–5039. [CrossRef]
10. Rocha, P.R.F.; Schlett, P.; Kintzel, U.; Mailänder, V.; Vandamme, L.K.J.; Zeck, G.; Gomes, H.L.; Biscarini, F.; De Leeuw, D.M. Electrochemical noise and impedance of Au electrode/electrolyte interfaces enabling extracellular detection of glioma cell populations. *Sci. Rep.* **2016**, *6*, 1–10. [CrossRef]
11. Medeiros, M.C.R.; Mestre, A.; Inácio, P.; Asgarif, S.; Araújo, I.M.; Hubbard, P.C.; Velez, Z.; Cancela, M.L.; Rocha, P.R.F.; de Leeuw, D.M.; et al. An electrical method to measure low-frequency collective and synchronized cell activity using extracellular electrodes. *Sens. Bio-Sens. Res.* **2016**, *10*, 1–8. [CrossRef]
12. Ribeiro, M.; Elghajiji, A.; Fraser, S.P.; Burke, Z.D.; Tosh, D.; Djamgoz, M.B.A.; Rocha, P.R.F. Human Breast Cancer Cells Demonstrate Electrical Excitability. *Front. Neurosci.* **2020**, *14*, 1–10. [CrossRef] [PubMed]
13. Cabello, M.; Ge, H.; Aracil, C.; Moschou, D.; Estrela, P.; Quero, J.M.; Pascu, S.I.; Rocha, P.R.F. Extracellular electrophysiology in the prostate cancer cell model PC-3. *Sensors* **2019**, *19*, 139. [CrossRef] [PubMed]
14. Moschou, D.; Tserepi, A. The lab-on-PCB approach: Tackling the μTAS commercial upscaling bottleneck. *Lab Chip* **2017**, *17*, 1388–1405. [CrossRef]
15. Gaetke, L.M.; Chow, C.K. Copper toxicity, oxidative stress, and antioxidant nutrients. *Toxicology* **2003**, *189*, 147–163. [CrossRef]
16. Searson, P.C.; Dawson, J.L. Analysis of Electrochemical Noise Generated by Corroding Electrodes under Open-Circuit Conditions. *J. Electrochem. Soc.* **1988**, *135*, 1908–1915. [CrossRef]
17. Yi, C.; Du, X.; Yang, Y.; Zhu, B.; Zhang, Z. Correlation between the corrosion rate and electrochemical noise energy of copper in chloride electrolyte. *RSC Adv.* **2018**, *8*, 19208–19212. [CrossRef]

18. Elveflow. Microfluidic Flow Controller (OB1 MK3+). 2019. Available online: https://www.elveflow.com/microfluidic-products/microfluidics-flow-control-systems/ob1-pressure-controller/ (accessed on 18 June 2021).
19. Elveflow. Microfluidic Flow Sensor 3 (MFS3). 2019. Available online: https://www.elveflow.com/microfluidic-products/microfluidics-flow-measurement-sensors/microfluidic-liquid-mass-flow-sensors/ (accessed on 18 June 2021).
20. Arjmandi, N.; Liu, C.; Van Roy, W.; Lagae, L.; Borghs, G. Method for flow measurement in microfluidic channels based on electrical impedance spectroscopy. *Microfluid. Nanofluid.* **2012**, *12*, 17–23. [CrossRef]
21. Perrier, R.; Pirog, A.; Jaffredo, M.; Gaitan, J.; Catargi, B.; Renaud, S.; Raoux, M.; Lang, J. Bioelectronic organ-based sensor for microfluidic real-time analysis of the demand in insulin. *Biosens. Bioelectron.* **2018**, *117*, 253–259. [CrossRef]
22. Pancrazio, J.J.; Gray, S.A.; Shubin, Y.S.; Kulagina, N.; Cuttino, D.S.; Shaffer, K.M.; Eisemann, K.; Curran, A.; Zim, B.; Gross, G.W.; et al. A portable microelectrode array recording system incorporating cultured neuronal networks for neurotoxin detection. *Biosens. Bioelectron.* **2003**, *18*, 1339–1347. [CrossRef]
23. Chung, B.G.; Choo, J. Microfluidic gradient platforms for controlling cellular behavior. *Electrophoresis* **2010**, *31*, 3014–3027. [CrossRef] [PubMed]
24. Walker, G.M.; Zeringue, H.C.; Beebe, D.J. Microenvironment design considerations for cellular scale studies. *Lab Chip* **2004**, *4*, 91–97. [CrossRef] [PubMed]
25. Young, E.W.K.; Beebe, D.J. Fundamentals of microfluidic cell culture in controlled microenvironments. *Chem. Soc. Rev.* **2010**, *39*, 1036–1048. [CrossRef] [PubMed]

 *micromachines*

MDPI

Review

Lab-on-PCB and Flow Driving: A Critical Review

Francisco Perdigones 

Electronic Engineering Department, Higher Technical School of Engineering, University of Seville, 41092 Seville, Spain; fperdigones@us.es

Abstract: Lab-on-PCB devices have been developed for many biomedical and biochemical applications. However, much work has to be done towards commercial applications. Even so, the research on devices of this kind is rapidly increasing. The reason for this lies in the great potential of lab-on-PCB devices to provide marketable devices. This review describes the active flow driving methods for lab-on-PCB devices, while commenting on their main characteristics. Among others, the methods described are the typical external impulsion devices, that is, syringe or peristaltic pumps; pressurized microchambers for precise displacement of liquid samples; electrowetting on dielectrics; and electroosmotic and phase-change-based flow driving, to name a few. In general, there is not a perfect method because all of them have drawbacks. The main problems with regard to marketable devices are the complex fabrication processes, the integration of many materials, the sealing process, and the use of many facilities for the PCB-chips. The larger the numbers of integrated sensors and actuators in the PCB-chip, the more complex the fabrication. In addition, the flow driving-integrated devices increase that difficulty. Moreover, the biological applications are demanding. They require transparency, biocompatibility, and specific ambient conditions. All the problems have to be solved when trying to reach repetitiveness and reliability, for both the fabrication process and the working of the lab-on-PCB, including the flow driving system.

Keywords: lab-on-PCB; microfluidics; flow driving; actuators; biomedical applications

Citation: Perdigones, F. Lab-on-PCB and Flow Driving: A Critical Review. *Micromachines* **2021**, *12*, 175. https://doi.org/10.3390/mi12020175

Received: 31 December 2020
Accepted: 5 February 2021
Published: 10 February 2021

1. Introduction

Lab-on-PCB has been the subject of increasing research over the last few years [1,2]. These devices emerged as a promising evolution of lab-on-chip devices [3–5] and the PCB-MEMS technology [6]. They share important properties with lab-on-chip devices—for example, small fluid volume and rapid response time. Particularly, the core of these devices is the integration of sensors for measuring the results of a reaction, and for controlling the parameters of the samples; and the integration of actuators for conditioning the samples and for moving those samples through the microfluidic platform.

The need for micromixing, microheating, and sensing in different parts of the microfluidic platforms makes the control of liquids mandatory. For this reason, the flow driving and fluid manipulation into a network of microchannels is one of the most important issues for lab-on-chip devices (LoC) and platforms [3,7], and particularly for lab-on-PCB. In this respect, the first attempts of developing a fluid manipulation date to late last century with a gas chromatographic air analyzer, and a miniaturized electrophoresis system [8,9]. Many works about flow driving have been carried out since those years, providing a large number of methods for performing similar tasks. Several of these methods have been integrated in lab-on-PCB: among others, pressurized microchambers, peristaltic pumps, and electrowetting on dielectrics.

Although lab-on-chip and lab-on-PCB have characteristics in common, and lab-on-PCB can be considered as a kind of lab-on-chip, they are different platforms. For example, unlike lab-on-chips, lab-on-PCBs are interesting due to the easy integration of microfluidics and electronics in the same platform, towards self-contained systems for microfluidic applications [10–12]. Apart from the integration, the interest in lab-on-PCB devices lies in the commercial availability of the PCB substrate with very reasonable dimensions at low cost [13,14]. Thanks to this characteristic, the lab-on-PCB devices can be disposable at low cost. This is important because the cleaning cost can be avoided. In fact, the cleaning implies an auxiliary and integrated microfluidic circuit or the use of the external facilities. The first option means an increase of the chip area, and thus a higher price. The cost of the second option is not worth worrying about because the devices are inexpensive. Moreover, the cleaning of small microchannels is very demanding, especially in biomedical applications. Therefore, the best choice to avoid cross-contamination from the biological and economical point of view is the use of freshly fabricated devices. As previously said, lab-on-PCB devices are a very interesting option due to their low cost. This is an important difference with respect to lab-on-chip devices, and makes the lab-on-PCB devices an attractive choice from the market point of view. However, these single-use devices imply environmental issues due to the metals of the PCBs. Fortunately, the electronics industry has been using PCBs for over 50 years and this issue is solved.

The main differentiating characteristics of lab-on-chip and lab-on-PCB are summarized in Table 1.

Table 1. Main differentiating characteristics of lab-on-chip and lab-on-PCB.

Characteristic	Lab-on-Chip	Lab-on-PCB
Materials for microfluidics	Silicon, glass, plastic	PCB, plastic
Substrate materials	Silicon, glass, plastic	PCB (rigid/flexible)
Maximum number of metal layers	2 (except silicon)	30 [15]
Fabrication of electronic tracks	Yes	Yes (low cost)
Commercially available substrate	Yes	Yes (low cost/integrated electronics)
Impact-resistant chips	Brittle (Silicon, glass)	Robust
Transparency	Yes	No
Highly integrated electronics	Yes (Silicon)	No
Discrete electronic components	SMD	SMD and through hole
Sensors/actuators integration	Yes	Yes (low cost)
Sensing performance	High	Medium
Biocompatibility	Yes	Yes (insulation layer)
Best application scenario	Optical and/or high sensitivity	The rest of the applications
Disposable at low cost	Yes (plastic)	Yes
Potential of commercialization	Low	Very high

As can be seen, the lab-on-PCB devices could be entirely fabricated using printed circuit boards; however, the lack of transparency makes them not very useful for optical measurement systems. In order to solve this issue, the PCB has been integrated with transparent materials, for example, glass, SU-8, PDMS, or kapton. For instance, the first platform integrating electronics and microfluidics using printed circuit boards (PCB) was developed by Jobst et al. [16] at the Technical University of Vienna in 1997, for the fabrication of a microdevice monitoring different metabolites by a biosensor array fabricated using glass. Two years later, Pagel and coworkers at the University of Rostock laid out the basis of the PCB technology, and they used it for developing several PCB-based devices, also named PCBMEMS devices [17–19]. The first lab-on-PCB itself also included a transparent cover. It was reported by Stefan Gassmann et al. [20] in 2007; see Figure 1.

Figure 1. The first lab-on-PCB reported by Stefan Gassmann et al. (Reprinted from [20], copyright (2007), with permission from Elsevier).

This device was composed of a microfluidic platform with integrated electronics, sensors, actuators, a transparent cover, and fluid manipulation for the detection of Fe^{3+}—that is, a lab-on-PCB with all its possible components and a specific application.

Glass, SU-8, PDMS, and kapton are very useful for fabricating prototypes, but there are better options with which to develop a commercial product. In this respect, an industrial lab-on-PCB device requires rapid mass production; that is, the fabrication of that product has to be performed at as low a cost as possible, while generating the largest number of products at the same time. For this reason, the thermoplastic materials are a good choice [21]. Most of them are transparent with a well-established mass production procedure, such as injection molding or hot embossing. These materials and fabrication methods should be chosen to fabricate a highly integrable flow driving systems, in order to develop marketable lab-on-PCB chips.

The final target in the development of lab-on-PCB devices lies on the mass production of commercial products. In fact, the fluid manipulation together with the PCB technology are very important from the point of view of the market, because they make it possible to tackle the development of inexpensive devices for many different biomedical and chemical analyses.

Nowadays, there is much work to do about the control of fluids and their integration into lab-on-PCB devices. Despite the improvements developed in recent years, lab-on-PCB is far from being robust. Unlike microelectronic chips, lab-on-PCB devices require a highly multidisciplinary R&D group, and they have a lack of standardization for both design and end-user interfaces.

Regarding the future outlook, the authors suggest the reading of [1], especially the Section 4, where a complete analysis of the future is performed.

Historically, the most used flow driving mechanisms in lab-on-chips, and especially in lab-on-PCBs are external pressure sources and syringe pumps. This is a well-established method, but reduces the portability of the whole system extremely. However, this method does not necessarily imply a disadvantage, as will be explained in the discussion section. The tendency to reduce or completely remove the connection between the microfluidic platform and the external sources is very positive from the portability point of view. Apart from the handling of fluids, the portability is hugely related to the marketable products. The reason for that lies in the commercial potential of the point of care devices, and the launch of new products into the market. The lab-on-PCB mechanisms mentioned in this review can seen in Figure 2.

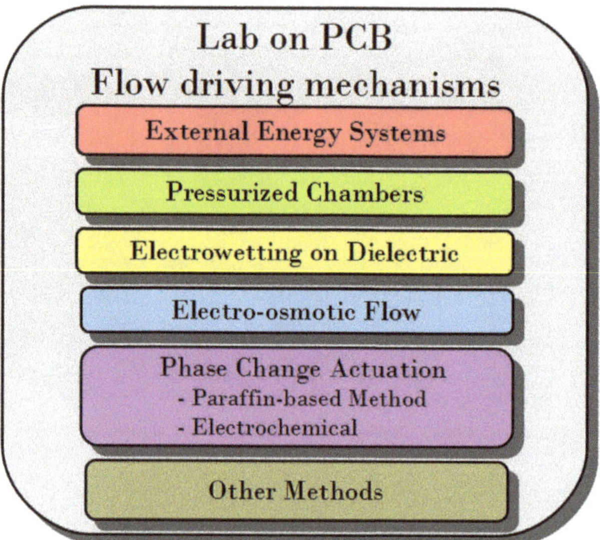

Figure 2. The external impulsion method together with the alternative ones mentioned in this review.

2. External Energy Systems

One of the most common methods to impulse liquids inside lab-on-chips and lab-on-PCBs consists of using syringe pumps. This method implies the use of tubing to connect these energy sources with the microfluidic platforms. In general, the number of pressure sources is a function of the number of liquids. This could be excessive for large-scale integration in microfluidics [22,23]. For example, a prototype of a PCB-based biosensor for rapid detection of *Salmonella* in food products [24] is shown in Figure 3. As can be seen, five tubes are connected to the device.

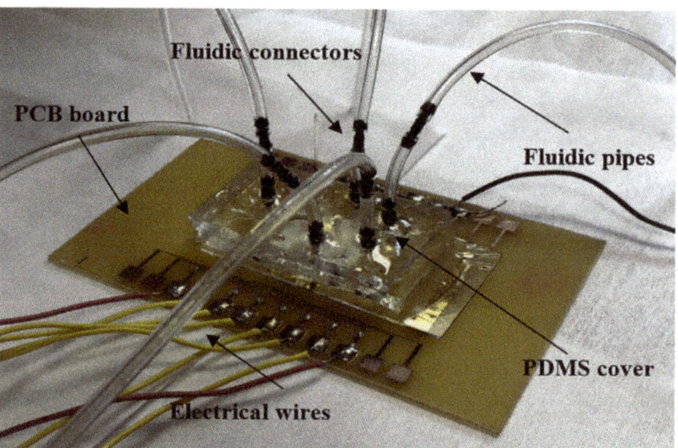

Figure 3. Prototype of a PCB-based biosensor for rapid detection of *Salmonella* in food products. The device requires the connection of five tubes, Ref. [24].

There are systems for multi-impulsion of liquids or gases to solve this problem, with the subsequent increase of cost. In addition, the connection of the tubes is an important issue to be solved. In this respect, multichannel chip-to-world interfaces for plug and play have been reported [25–27]. This solution consists of defining ports in microfluidics which tackle the required standardization. However, the portability continues to be an issue to face.

Many lab-on-PCB devices use external energy sources, such as syringe pumps, for moving the liquids with continuous flow. Although several combinations of external energy sources with lab-on-PCBs have been reported in the past [16,28–30], the examples chosen for this mechanism are a representative group of the recent past and present. For instance, the lab-on-PCB device reported by Moschou et al. [31] (2015) demonstrated the integration of stable Ag/AgCl pseudo-reference electrodes. In order to do so, the authors used a laboratory syringe pump (Chemyx Inc., Fusion 200, Stafford, TX, USA) for moving a buffer continuously through the reference electrode over 24 h. In 2017, a lab-on-PCB-based cytometer for detecting circulating tumor cells and enumeration was developed [32]. In this work, the biological samples were driven by a syringe pump (PHD 22/2000, Harvard Apparatus, MA, USA).

The PCB-based microfluidic platform for electrochemical detection of cancer biomarkers reported in [33] (2016) and the lab-on-PCB for conditioning the medium of cell cultures and mixing fluids described in [34] (2017), also require a connection to a syringe pump.

Recently, the PCB-based thermocycler for PCR developed in [35] (2019) required a syringe pump in order to move the DNA sample through a microchannel; the lab-on-PCB reported in [36] (2019) for organotypic cultures required continuous flow. Therefore, it needed to be connected to a external source (New Era Pump Systems, Inc, Farmingdale, NY, USA) to feed the tissues with culture medium. In addition, the lab-on-PCB for rapid and high sensitivity DNA quantification reported in [37] (2019) also made use of a syringe pump (Cole Palmer 230-CE) for continuous flow experiments; the reagents were delivered into the lab-on-PCB inlet; see Figure 4. The same syringe was used for delivering glucose samples in a lab-on-PCB for electrochemical glucose sensing (2020) [38]. Finally, peristaltic micropumps have also been used as external impulsion for lab-on-PCBs [39].

As can be seen, the external sources have been used for moving liquids in lab-on-PCBs from the beginning of this kind of device until today. This fact shows that this method continues being a good alternative to developing new devices with new applications. Although the external syringe pumps reduce the portability of the system, they are an interesting choice to demonstrate the integration of sensors and actuators in lab-on-PCB. Apart from this, many studies has been performed in order to remove the connection tubes from the microfluidic devices, aiming for portability.

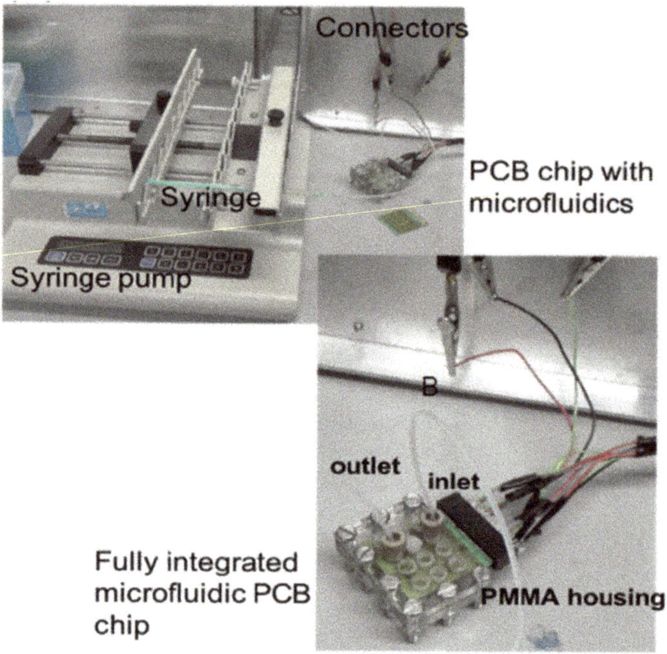

Figure 4. Lab-on-PCB for rapid and high sensitivity DNA quantification. The experiments are performed using an external syringe pump (Reprinted from [37], copyright (2019), with permission from Elsevier).

3. Pressurized Chambers

The method based on pressurized microchambers [40,41] is a good alternative for moving liquids in lab-on-PCBs. This is a mechanism which allows the storage of pneumatic energy in a microchamber of the lab-on-PCB, so that the connection tubes can be avoided. Once the energy is available, it can be released by the opening of a microvalve. The activation of those microvalves is electrically performed. In order to do so, a gold wire with a diameter of 25 μm is used as a microheater.

The pneumatic energy is stored inside SU-8 microchambers as high pressure air. The releasing of the air is achieved by the destruction of a thin vertical wall due to both the increase of temperature and the pressure of the microchamber; see Figure 5A. The gold wire is perpendicular to the wall and so it is not optimal. In this respect, several microvalves with the wire completely embedded in the wall have been reported [42–44]. These devices have vertical walls except the ones described in [43,45], where the SU-8 wall was a membrane transferred to the PCB substrate; see Figure 5B.

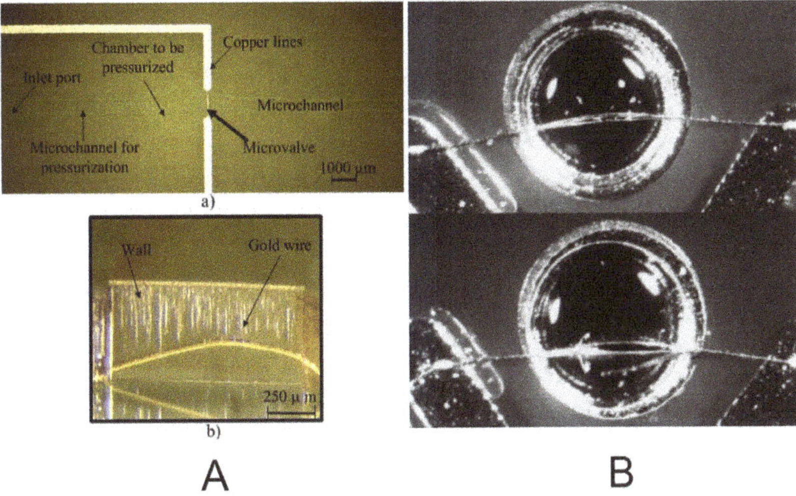

Figure 5. (**A**) Pressurized microchamber with vertical walls and an embedded gold wire (copyright (2014) IEEE. Reprinted, with permission, from [44]). (**B**) Pressurized microchamber composed of a planar SU-8 membrane with a completely embedded gold wire (Reprinted from [43], copyright (2010), with permission from Elsevier).

These microvalves were improved by using thin copper lines instead of wires [40,46]. They were fabricated by wet etching at the same time as the copper electronics tracks. This improvement implies the removing of the wire bonding step, so that the gold and the facilities involved to perform the bonding are avoided. In this case, the microvalve is used as a fuse, in order to destroy the SU-8 wall. An example of these devices is shown in Figure 6.

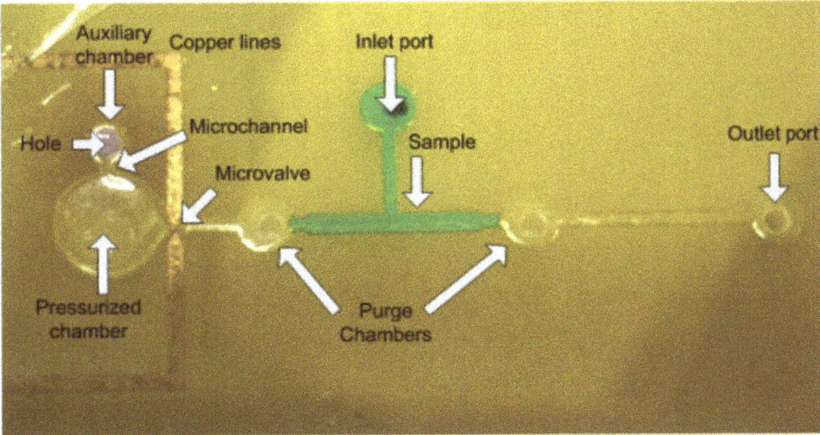

Figure 6. Impulsion system based on an SU-8 pressurized chamber and a copper line fuse. (Reprinted from [40], copyright (2015), with permission from Elsevier).

Several lab-on-PCB devices have been fabricated using this method—for example, the lab-on-PCB micromixer reported in [47], and the prototype developed to demonstrate the integration of a protocol in a lab-on-PCB platform [48].

Lab-on-PCBs have been fabricated using a thermoplastic (PMMA) and PCB, that is, materials compatible with mass production and multiple processes; see Figure 7. This device is closer to a marketable product than those using SU-8-based materials. Actually, the device shown in Figure 7 is fabricated via a computer numerical control (CNC) machine. This is not a mass production procedure, but the PMMA can be processed by injection molding or hot embossing.

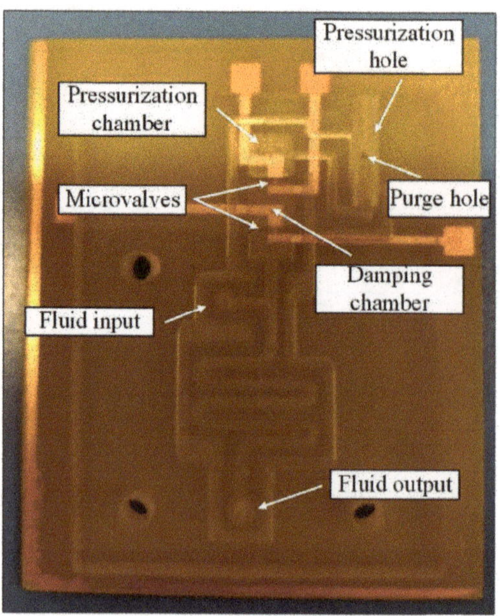

Figure 7. Pressurized chamber fabricated using a thermoplastic and PCB, with copper lines as a fuse [49]. (Copyright (2018) IEEE. Reprinted, with permission, from [49]).

In this case, the air has to be inserted in the microchamber using plungers [49]. The destruction of the fuse opens a small microchannel through which the air leaves the microchamber to impulse the liquid stored in the microfluidic circuit.

This method is mechanically simple due to the impulsion being based on no moving parts. With the electric energy required to destroy the copper fuse, the control and signal processing can be included in a unique electronic circuit.

Regarding the limitations of the method, it is important to comment on the lack of biocompatibility due to the destruction of the copper fuse. The debris of the copper lines could damage the biological samples. The authors propose an inert and intermediate liquid between the pressurized microchamber and the rest of the microfluidic circuit, so that the samples will never be in contact with the contaminated air. Nevertheless, the integration of the inert liquid reduces the simplicity of the whole system. In addition, it increases the complexity of fabrication. These facts imply an increase of the fabrication cost as a commercial product.

Finally, the gas permeability of the polymeric wall is important because the pressurized microchamber could discharge in the long-term. This is a problem for long-term storage as a product. Moreover, the handling of fluids using this method is limited to small samples. Continuous flow is not possible to achieve; for instance, the insertion of medium in a long-term cell culture cannot be managed by pressurized chambers. This is not a problem because this method is not intended to do so. There are better choices to those applications, for example, external energy sources.

4. Electrowetting on Dielectric

Electrowetting on dielectrics (EWOD) is one of the typical techniques of digital microfluidics [50–52]. This is a method whereby an electric field changes the wetting of a droplet, in contact with insulated and hydrophobic electrodes. The droplets are placed between two parallel layers; the bottom one is the substrate, which includes an electrode array covered with thin dielectric layer, and the top layer could be either a passive top plate or a ground plate. This method consists of switching the voltage to electrodes, so that the surface tension gradient can be modified, generating asymmetric contact angles, and the subsequent driving forces. Thus, the droplet is moved in two dimensions.

Among others, the advantages of EWOD devices include flexible device geometry, compatibility with other technologies, and simple electronic instrumentation. In addition, this technique allows easy manipulation of several reagents at a time, with a reduction of reagent volume, a short analysis time, and high sensitivity. Despite these advantages, the biomolecular adsorption of biological material due to the hydrophobic layer, together with electrolysis and evaporation of the small volume of liquids, and cross-contamination are the drawbacks of the technique, to name a few.

The PCB technology provides several advantages to EWOD devices. The typical fabrication process of these devices implies expensive clean rooms when processing glass substrates, resulting in smooth surfaces with reliable motion of droplets at voltages below 100 V. Integrated circuits offer the smooth surface topography required in the droplet manipulation. However, the PCB substrates provide low cost fabrication and quick turnaround time with slightly higher driven voltages [53].

Many studies have taken advantage of the up-sides. For instance, the platform reported in [54] demonstrates the combination of lab-on-PCB and EWOD for two-plate and one-plane devices. The two-plate fabricated devices were used for moving, dispensing, merging, and splitting droplets in the range of 150–300 nL. On the other hand, the one-plate devices managed droplets with volumes of up to 3 µL. In addition, Parylene-C and PDMS were used as dielectric layers and the voltages ranged between 300 and 500 V at 18 kHz. Similar platforms have been reported for biological application and lower driven voltages, about AC voltage of 200 V at 1 kHz, for moving droplets of about 30 µL [55]. Besides, the works reported in [56] moved droplets of about 10 µL with a speed of 3 mm/s by applying a high DC voltage, 400 V. This very low cost device is composed of a planar array of six electrodes with silicone rubber as the dielectric layer and a commercially available water repellent as the hydrophobic layer. A similar open platform was reported in [57]; a planar array of 64 electrodes was developed. In this work, silicone oil and Parafilm M were used as a dielectric hydrophobic layer. The speed was 15 mm/s for droplets with a volume of 1850 µL, using a voltage output frequency of 10 Hz. This device is shown in Figure 8, where the open EWOD platform and the electronic circuit to control the droplets can be seen. Unlike the previously mentioned devices, the one reported in [58] was a two plate system with an array of 24 × 24 electrodes controlled by seven control signals at 15–30 V. It was composed of a dielectric layer (SU-8) and a hydrophobic layer (Teflon AF1600) deposited on the PCB.

Figure 8. The open EWOD platform and the electronic circuit to control the droplets can be seen [57].

The work described in [53] scales the PCB-based electrowetting devices to larger arrays, with smooth surface topography if taking into account the roughness of the PCB substrate. The authors state that they successfully moved, merged, and mixed droplet with volumes from 2 to 1200 μL. On the other hand, multilayer PCB substrates have also been used for EWOD [59–61]. Finally, PCB has also been used for contactless electrowetting, that is, modifying the contact angle by air ionization [62].

Apart from these prototypes and studies, the development towards a commercial product reported in [63,64] proposes droplet-based pyrosequencing and point of care devices as lab-on-PCB; see Figure 9.

In particular, the device is intended to perform both immunoassays for cardiac troponin I and real-time PCR assays with 300 nL droplets. Besides, the fabrication process is performed by utilizing mass production techniques.

As can be seen, lab-on-PCB devices based on electrowetting on dielectrics are a very interesting choice for many biological applications [65]. The technology is mature enough to stimulate the creation of companies based on PCB for biomedical applications [66].

In general, the problem with this kind of platform is related with the number of electrodes and the electronic circuit; that is, a large number of electrodes implies a larger number of electronic components in the circuit. Even so, this fact does not seem to be a drawback for industrial production. Regarding the fabrication process, it only requires the commercially available printed circuit board process and several coatings, together with a plastic part to store liquids. Finally, it is worth mentioning that this method does not necessarily require microchannels, and thus the related fabrication processes.

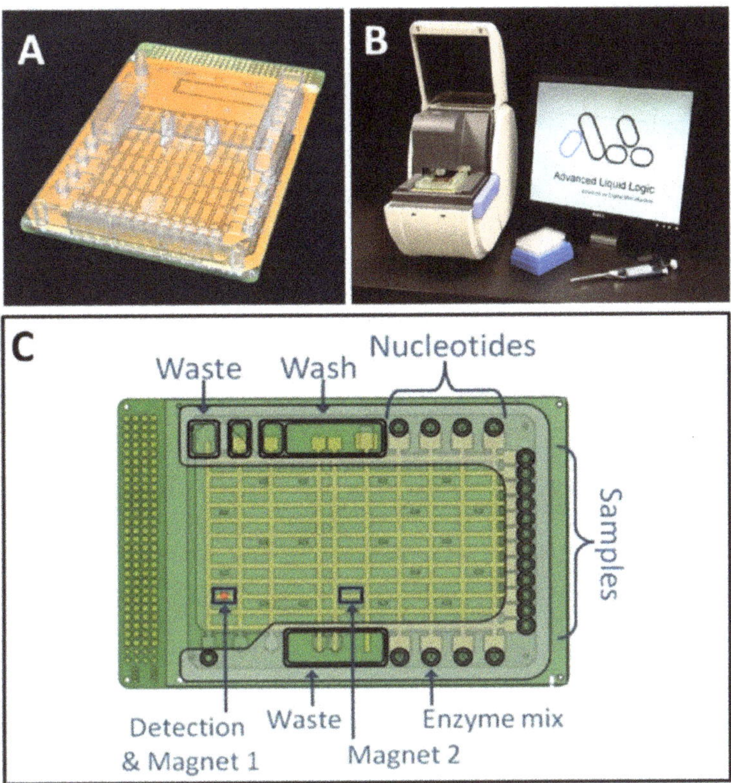

Figure 9. (**A**) Assembled multiwell-plate-sized PCB-based cartridge. (**B**) Photograph of the control instrument. and (**C**) Sketch of the cartridge showing the locations of sample and reagent wells (Reprinted with permission from [63], copyright (2011), American Chemical Society).

5. Electroosmotic Flow

This kind of pump is based on an electrokinetic phenomenon (electroosmosis) [52,67,68]. This process is used for impulsion liquids through microchannels or porous media, via the application of an external electric field. The electroosmotic effect takes place when fluids containing polar molecules are in contact with solid surfaces, so that electric charges appear in those surfaces which are in contact with the liquids. Those charges are negative when the working fluid is water in contact with an insulating solid. In addition, positive charges are induced at the same time in a thin layer of the water very close to the surface, in order to keep the electric neutrality. The layers of electrical charges of the fluid are named "Helmholtz layers", and the combination of these positive and negative layers is called the "electrical double layer". When an electrical field is applied inside the microchannel, forces appear on the positive layer and the charges are moved towards the negative electrode, transporting the liquid with them. The majority of electroosmotic pumps worked under direct current voltage. However, bubbles could appear at the driven electrodes due to the electrolysis. In this respect, the use of low alternating voltage can reduce or even eliminate the generation of bubbles .

Among other things, the main advantage of electroosmotic pumps is the generation pulse-free flows. In addition, the flow magnitude and direction can change instantly. Finally, like the impulsion systems based on pressurized chambers, electroosmotic pumps have no moving parts. Regarding the limitations, electrolysis and the bubble generation could appear at the metal electrodes. Moreover, the flow rates are low if compared to other integrated micropumps, and the microchannels required to achieved a given flow rate are relatively narrow.

Printed circuit boards are a good candidate to develop electroosmotic pumps due to their metal layer. This layer is used for fabricating the driven electrodes with the typical photolithographic process. Consequently, the integration can be performed with ease. The rest of the electroosmotic pump has to be fabricated with a different material.

Several electroosmotic pumps have been integrated on PCBs—for example, the prototype reported in [69], which was fabricated using a PCB and SU-8; the Flame Retardant 4 (FR4) was the substrate, SU-8 was the material chosen with which to fabricate the microchannels and microchambers, and the copper layer was used for building the electrodes, electrical connections, and pads. The device is inexpensive but the materials make the mass production difficult. A similar pump with the same fabrication process was reported in [70]; see Figure 10. This pump is able to provide a flow rate of 1 μL/min at a direct current voltage of 60 V.

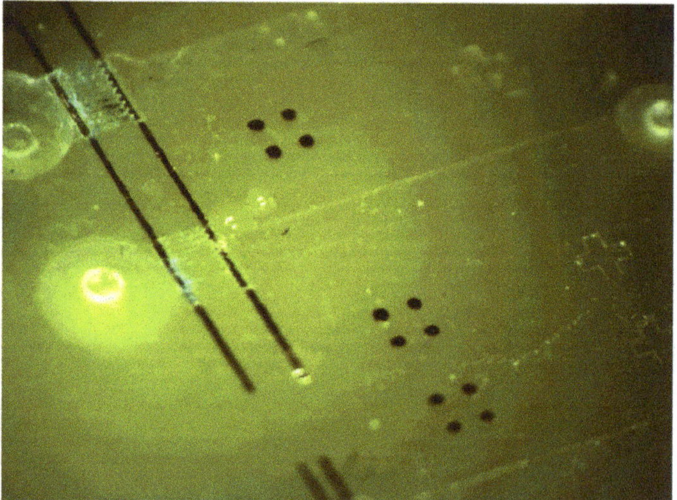

Figure 10. Prototype of an electroosmotic micropump for lab-on-PCBs. (Copyright (2012) IEEE. Reprinted, with permission, from [70]).

A different material was integrated with the PCB in a study presented in [71]; the authors describe a lab-on-PCB compatible with this flow driving method. The fabrication materials include a dry resist and PCB, and require hot pressing and several photolitographic processes. In addition, it includes surface mounted electronic components (SMD) intended to take measurements; for example, the authors performed optical experiments via the integration of two embedded SMD blue LEDs. They also integrated a temperature sensor and a resistor.

The work reported in [72] shows the design and characterization of a passive, disposable wireless for lab-on-PCB for particle and fluid manipulation. Unlike the previously mentioned devices, this one was fabricated on a flexible PCB (lab-on-a-film) integrating a receiving coil, an array of interdigitated electrodes (IDE), and two SMD components, a diode and a capacitor. It works at low voltages, and can perform three microfluidic operations depending on the wirelessly-controlled voltage, so that when the signal over an array of interdigitated electrodes is about 0.7 V, the IDE performs particle enrichment. The IDE works as an active mixer at 2 V; and as an AC electroosmotic pump when the voltage is 3 V. All of these functions are performed by the device itself with overall dimensions of 10×20 mm^2; see Figure 11.

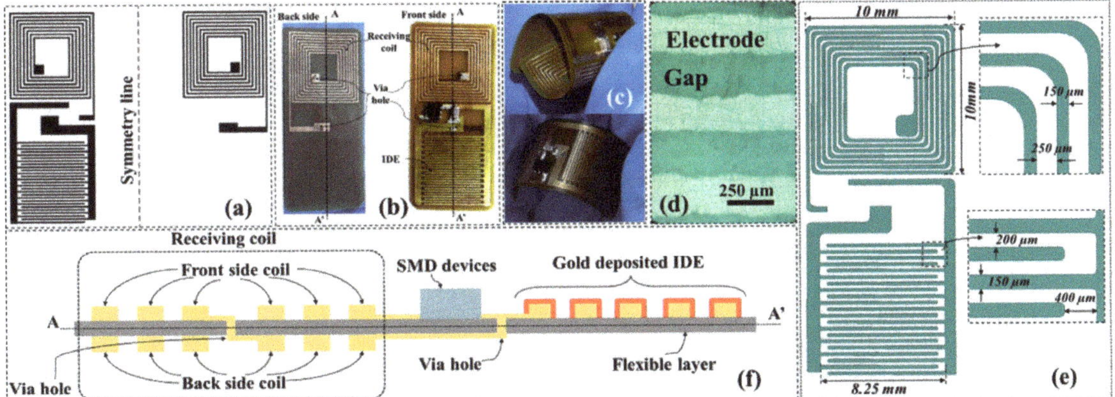

Figure 11. (**a**) Layout used for the device fabrication, (**b**) fabricated biased-AC electroosmotic lab-on-a-film pad, (**c**) demonstration of the device's flexibility, (**d**) detail of the metal traces, (**e**) dimensions of the device pad, (**f**) cross-sectional view of the device along AA′ cut (Reprinted from [72], copyright (2016), with permission from Elsevier).

Finally, the micropump reported in [73] can be surface mounted on a commercial PCB. This pump is composed of a nanoporous membrane to provide an electroosmotic flow for a maximum flow rate of 8 µL/s at a voltage of 2 V.

Summarizing, electroosmotic pumps have been integrated into lab-on-PCBs for fluid manipulation. They provide a low cost choice for electrically controlled pulse-free continuous flows compatible with the integration of sensor and actuators. Nevertheless, external or integrated liquid reservoirs and narrow microchannels are required. In addition, electrolysis and bubbles could appear with inappropriate activation. On the other hand, the working of these devices with copper electrodes implies oxidation and the subsequent contamination of the fluid, and eventually the blocking of the microchannels (see the electrodes of the device of Figure 10). In order to avoid the oxidation, additional fabrication steps are required so that the fabrication cost increases due to both new facilities and new fabrication materials. In addition, the narrow microchannels are difficult to fabricate with mass production equipment with reasonable tolerance at low cost. This means a drawback from the industrial point of view.

6. Phase Change Actuation

The flow driving systems based on phase changes, such as microvalves and micropumps, have been used for microfluidic handling. The main methods for this purpose are based on paraffin [74] and electrochemical micropumps.

The paraffin-based method makes use of the thermal properties of the paraffin to change its phase from solid to liquid and vice versa. This material modifies its volume as a function of the temperature so that the higher the temperature is, the more the paraffin increases its volume, typically about 10–15%. This increase of volume and the melting are used for opening and closing valves in order to modify a flow rate. These systems have been included in lab-on-PCB platforms due to their advantages, such as simple design, and more importantly, the low cost of the paraffin. In addition, the paraffin-based valves can be activated several times, and they support high pressures [75]. However, these devices have several drawbacks, for example, the integration of the paraffin. Moreover, the characteristic of the material makes the complete system temperature-dependent. Finally, the time response to manipulate the liquids is high, about tens of seconds.

The increase of temperature required to change the phase of the paraffin has to be performed using a thermal actuator, that is, a microheater. Thus, the copper layer of the PCB substrate is very suitable to fabricate that microheater. It is developed as a serpentine copper line whose heating is governed by the Joule effect. This fact makes the paraffin-based actuation method easy to integrate in lab-on-PCB.

The paraffin-based method for fluid manipulation has been integrated in several lab-on-PCBs. Bodén et al. [76], in 2008, developed an on-chip liquid storage and dispensing for lab-on-PCB. The system has three integrated liquid reservoirs and dispenser units for 10 μL sample volumes. Those reservoirs can store either reagents or liquids for buffering or rinsing. The device is composed of a PCB substrate with an integrated copper microheater and an epoxy structure for the microfluidic circuit. Once the microheater is activated, the paraffin melts and expands, increasing the pressure. Therefore, the fluid is driven toward the microchambers and microchannels.

Recently, in 2019, Wang et al. [77] reported an on-board control of paraffin-based microfluidics manipulation on an active centrifugal lab-on-PCB for plasmid DNA extraction. The impulsion is based on both centrifugal forces and heating. The device is composed of a plastic microfluidic layer fabricated using a PMMA master, and a commercial PCB layer as a substrate. The heating is achieved by resistors in a PCB substrate. Unlike the previously mentioned paraffin-based lab-on-PCBs, the resistors are not serpentine-shaped copper lines. In this case, the temperature is raised using SMD resistors as thermal actuators. This implies a lower area of heating compared with the copper lines. Once the paraffin is melted, the centrifugal forces drive the fluid.

Regarding the electrochemical method, the most representative one is based on electrolysis, that is, the gas generation inside a liquid. Typically, the chosen liquid is water because it is inexpensive and easy to work with, and the resulting gases are hydrogen and oxygen. This method was applied to a PCB-based micropump with the maximum flow rate of 31.6 mL/min and maximum backpressure of 547 kPa (at 34 μL/min) [78,79]. The water electrolysis is achieved using integrated interdigitated microelectrodes on the PCB, where the metal of the electrodes is fabricated using the copper layer of the PCB with electroplated gold; see Figure 12.

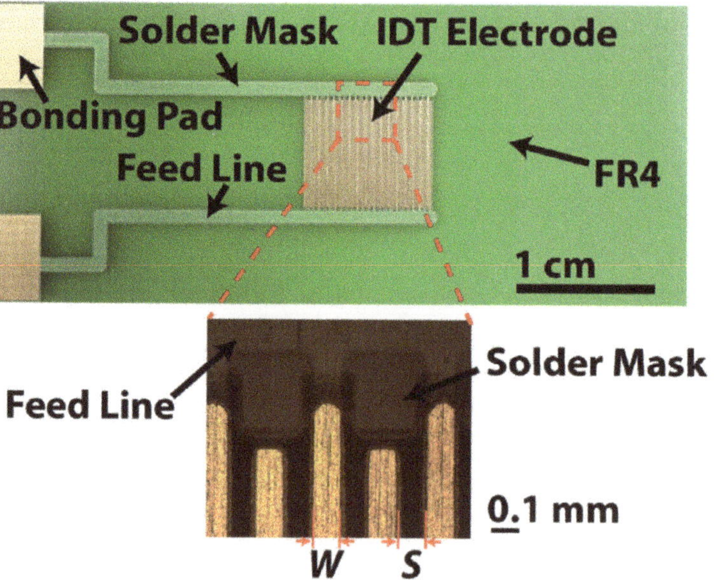

Figure 12. Electrochemical PCB-based impulsion chip with enlarged detail of the microelectrode fingers (Reprinted from [78], copyright (2018), with permission from Elsevier).

A similar method was reported in [80] with a flow rate ranging from 20 μL/min to 135 μL/min. This device also uses integrated metal electrodes to perform the electrolysis of water. However, the electrodes are made of copper without electroplated gold. The very well-finished lab-on-PCB reported in [81] also uses the electrolytic gas generation inside a hydrogel. These electrolytic and PCB-based impulsion systems require the precise integration of a material inside the system, especially the last one, where the hydrogel has to be dispensed into the microchambers. Finally, the electrolytic method was applied for flow driving in a USB-based lab-on-PCB device for microparticle generation [82].

Finally, the self-contained, fully integrated lab-on-PCB for sample preparation, PCR amplification, and DNA detection [11] is one of the most representative lab-on-PCB which uses, among other things, electrochemical flow driving. The device includes microchannels, microchambers, micromixers, microvalves, micropumps, and microheaters. The chip is able to manipulate initial samples in the order of microliters and milliliters. In order to do so, electrochemical micropumps are used for driving milliliter volumes. It is based on the electrolysis of water between two platinum electrodes to generate gases when a current is applied. These gases increase the pressure that moves liquids in the device. In addition, a thermopneumatic pump for driving microliter volumes is used. The expansion of the gas is performed in a chamber attached to a PCB-based and resistive microheater. The resulting air expansion drives the liquids into the microchannels and microchambers of the device. Moreover, this device also integrates paraffin-based microvalves for fluid manipulation in lab-on-PCB; see Figure 13. The same authors from Motorola Labs used these actuators for DNA amplification, Ref. [83]. The device was composed of a PCR chamber and paraffin-based microvalves. *Escherichia coli* K12 cells were used in the experiments. Regarding the main fabrication materials, the device was built using mass production materials, that is, polycarbonate and a PCB as the substrate. The fluid manipulation was performed by resistive microheaters to provide the thermal actuation to the paraffin.

Summarizing, the mentioned paraffin-based lab-on-PCBs are robust platforms for microfluidic handling. The reported results are promising, as can be deducted from the reported biomedical applications. Nevertheless, the integration of the paraffin implies complex fabrication steps which go against the low cost mass production, and the subsequent competitiveness as a commercial product. The electrochemical methods are also robust but they have the same drawback as the paraffin-based one, that is, the integration of a small quantities of materials—in this case, water or hydrogels.

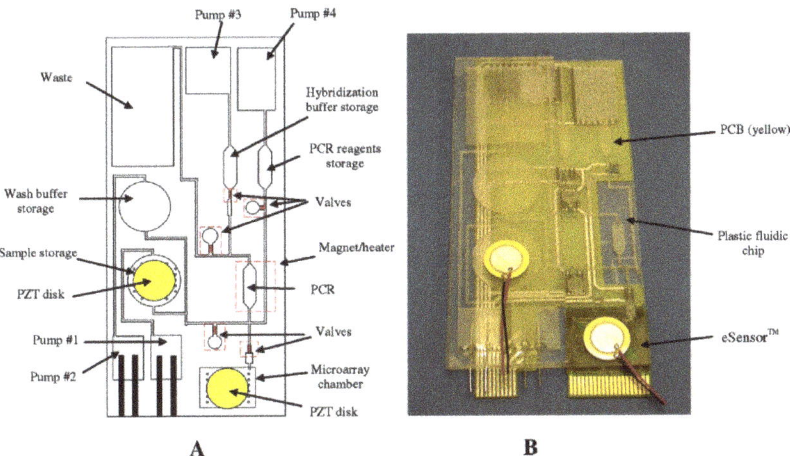

Figure 13. Self-contained, fully integrated lab-on-PCB for sample preparation, PCR amplification, and DNA detection. (**A**) Sketch of the plastic fluidic lab-on-PCB. Micropumps 1–3 are electrochemical, and pump 4 is thermopneumatic. (**B**) The integrated lab-on-PCB consists of a plastic microfluidic chip, a printed circuit board (PCB) substrate, and a Motorola eSensor microarray chip (Reprinted with permission from [11], copyright (2004), American Chemical Society).

7. Other Methods

Apart from the previously mentioned methods, there are additional procedures related to flow driving on lab-on-PCBs. These methods have been reported to a lesser extent, but they are worthy of being mentioned. The methods mentioned in this section are peristaltic PCB-based impulsion, passive microfluidics, and several lab-on-PCB compatible methods.

The microfluidic peristaltic pumps are based on microchannels' deformation to move small volumes of fluid from the inlet towards the outlet, achieving a net continuous flow [84,85]. In order to do so, an electronic AC signal is used for driving the fluid. These micropumps can easily change the direction of the flow. In addition, the flow rate and pressure can be modified by changing the peak-to-peak voltage values, the frequency, and the phase difference of the driving voltages. Nevertheless, they have several drawbacks, such as moving parts and large working area on the substrate.

The integration of peristaltic pumps into printed circuit boards is not easy. This task implies the development of actuators with moving parts to achieve the peristaltic effect. This could be the reason for the lack of pumps of this kind for lab-on-PCBs, if compared with other impulsion methods. However, there are several approaches of PCB-based peristaltic micropumps for lab-on-PCBs. One of these devices was reported in [86,87]. These pumps were composed of a printed circuit board and piezoelectric commercial actuators (buzzers in electronic equipment). The experimental results show a maximum working frequency of 50 Hz with a maximum speed of 3 mm/min for a driven voltage of 140 V. The same working principle was used in the device described in [88]. In this case, the maximum average flow rate was 500 µL/min when the peak-to-peak driven voltage was 100 V at 10 Hz, the maximum backpressure being 760 Pa. The dimensions of the PCB substrate when the pump is integrated were about 10×6 cm^2. This seems to be large for a integrated pump, but microfluidic circuits and sensors could be included in the remaining area, or even on the opposite side of the PCB.

As a conclusion, it is important to highlight that the main drawback is the required moving parts and its integration into the PCB substrate. That integration implies increasing both the complexity of the fabrication process and the dimensions of the device. Although the PCB substrate makes the device cheaper, the mentioned disadvantages go against the target of lab-on-PCB devices, that is, the inexpensive mass production focused on the market. In fact, up to now, the characteristics of the PCB substrate have not shown interesting advantages in the fabrication of these pumps. Unlike the systems for fluid manipulation mentioned in previous sections, in peristaltic micropumps the copper layer cannot be used for developing the actuator in charge of the liquid movement. In addition, they need external or integrated liquid reservoirs.

The surface acoustic wave (SAW) methods for manipulating fluids are commonly used in microfluidics and lab-on-chip [52,89]. The development of this kind of device using printed circuit boards is not easy due to the substrate material having to be piezoelectric. The typical application of PCBs to these devices is the connection between the radio frequency signal generator and the SAW actuator [90]. However, the microelectrodes can be developed using the copper layer of a PCB substrate. In this case, these electrodes have to be clamped to a piezoelectric substrate. This solution was reported in [91] for fluid and cell manipulations with maximum droplet velocities of about 40 mm/s.

The PCB-based liquid dispenser reported in [92] is a microfluidic pump composed of a oscillating membrane made from a flexible PCB, and magnetically actuated via Lorentz force. Following with the same application, a PCB-based dispenser for biomedical integrates a bending CNT actuator into a PCB, which enables the induction of movement [93]. In addition, a thermally actuated one-shot liquid dispenser, with an actuation based on highly expandable microspheres on PCB has been reported [94].

On the other hand, passive microfluidics is other typical and well-known method used for driving fluids [95]. This method has also been used for printed circuit boards to develop high-performance PCB-based capillary pumps for affordable point-of-care diagnostics [96,97]. The flow can be created without actuation; for example, take the PCB-based self-breathing fuel cells [98,99]. The working itself leads the motion of the liquid. Despite all this, the handling of fluids is one of the most important issues for lab-on-chips and lab-on-PCBs.

There are interesting flow driving methods compatible with lab-on-PCB devices, but they have been developed using different substrates. The first one is based on the integration of thermo-expanding microspheres [100] into silicon substrates. These microspheres were also used for blocking the flow in microchannels. The second one is the impulsion system reported in [101,102] which uses azobis-isobutyronitrile (AIBN) as the solid chemical propellant and a gold microheater. This device is composed of a cyclic oleofin copolymer substrate and gold as a metal. The AIBN is heated at 70 °C to produce nitrogen gas in order to impulse the liquid samples. Finally, dielectrowetting is an interesting compatible technique which creates stronger wetting forces than EWOD [103]. Droplets can be created, transported, splitted, and merged using this technique [104].

8. Discussion and Conclusions

The use of the copper layers of the PCB to develop actuators has provided good results. Several PCB-based actuators are easily integrable using that layer, for example, microheaters, microvalves, and electrodes. These actuators have successfully been used for driving flow on lab-on-PCBs. The level of development in the majority of them is suitable. However, the integration of these actuators has drawbacks. The first one is the increase of the area of the substrate due to the integration of the actuator. This fact means an increase of the cost of the final product, especially for single use devices. In this respect, the use of external impulsion devices for lab-on-PCBs implies smaller, simpler and cheaper PCB-chips. This is important because the PCB-chips are the most important source of benefits, especially if they are disposable. In addition, the use of external impulsion devices makes the quality control of the flow driving easy. The reason for this characteristic lies in the use of a unique device for driving fluids for many lab-on-PCBs, instead of a new integrated device for each lab-on-PCBs. Therefore, the use of external impulsion devices is a good choice for no portable lab-on-PCB devices. One drawback of these devices is the change of the tubing for each experiment due to the contamination. This trend could suggest focusing the research on the miniaturization of external devices with no integration so that the portability is applied to the whole system, that is, the electronic control device with small external impulsion devices, and more importantly, small disposable PCB chips. This implies the development of well-defined and standard fluidic and mechanical ports.

It is important to mention that the flow driven devices are not always necessary for lab-on-PCBs. Examples of these devices include the wearable detection biomedical device for real-time detection of glucose in sweat reported in [105] and a PCB-Based thermocycler for PCR [106]. These devices do not require the motion of fluids.

Table 2 summarizes the lab-on-PCB devices with the flow-driving active mechanisms mentioned in this review, except the external energy sources. The table shows the flow driving method and the actuator, the fabrication materials additional to the PCB substrate, the nature of the flow, and the application of the lab-on-PCB. In addition, Table 3 shows a quantitative comparison.

This review described the active flow driving methods for lab-on-PCB devices, mentioning their characteristics from the industrial point of view. In general, there is no perfect method because all of them have drawbacks.

The main problem with regard to marketable devices is the complex fabrication due to the integration of additional fabrication material. In addition, the method of assembly of the LoPs increases that difficulty.

Moreover, the biomedical and biochemical applications are very demanding. Among other things, they require transparency, biocompatibility, and specific ambient conditions, for example, temperature, humidity, or pH. Particularly, these requirements have to be compatible with the flow driving method, together with the rest of the lab-on-PCB, including the sensing. All the problems have to be solved to provide consistency and reliability for both the fabrication process and the function. In this respect, the most developed method for commercial PCB-based flow driving applications, which better fulfills those requirements, is electrowetting on dielectrics. Nevertheless, it has drawbacks which could be solved with the combination of EWOD with a different flow driving method.

As can be seen in this review, lab-on-PCB devices have been developed for many biomedical and biochemical applications. However, much work has to be done towards commercial applications. This is especially challenging if taking into account the boundary conditions of the market, that is, repeatability, reliability, and acceptable fabrication processes. Even so, the research of devices of this kind is rapidly increasing, so interesting solutions are being developed day to day. The reason for this lies in the great potential of lab-on-PCB devices to provide marketable devices.

Table 2. Characteristics of active microfluidic, pressure-driven lab-on-PCBs.

Method	Actuator	Materials	Flow	Application	References
Pressurized	valve	SU-8	Samples	General Purpose	[40,44,46,48]
Pressurized	valve	SU-8/Au	Samples	General Purpose	[42,47]
Pressurized	valve	PMMA	Samples	General Purpose	[49]
EWOD	Electrodes	Dielectric	Droplet	General Purpose	[53,55,56,62]
EWOD	Electrodes	Glass/dielectric	Droplet	General Purpose	[54,59–61]
EWOD	Electrodes	Polymer/dielectric	Droplet	Clinical diagnosis	[64]
EWOD	Electrodes	SU-8/Teflon	Droplet	General Purpose	[58]
Osmotic	Electrodes	SU-8	Continuous	General Purpose	[69,70]
Osmotic	Electrodes	1002F	Continuous	General Purpose	[71]
Osmotic	Electrodes	Gold	Continuous	General Purpose	[72]
Osmotic	Electrodes	AAO/PDMS	Continuous	General Purpose	[73]
Peristaltic	Piezo-disc	Brass	Continuous	General Purpose	[86–88]
Paraffin	Heater	Paraffin/Epoxy	Samples	General Purpose	[76]
Paraffin	Heater	Paraffin/PDMS	Samples	DNA extraction	[77]
Thermoneumatic	Heater	Paraffin/PC	Samples	DNA amplification	[11]
Electrolytic	Heater	Paraffin/PC	Samples	DNA amplification	[11]
Electrolytic	Electrodes	PMMA	Continuous	General Purpose	[78,79]
Electrolytic	Electrodes	Polymer/Pt	Continuous	General Purpose	[80]
Electrolytic	Electrodes	Polymer	Continuous	PCR and Analysis	[81]
Electrolytic	Electrodes	Glass	Continuous	Droplet generation	[82]
SAW	Electrodes	$LiNbO_3$	Droplet	Cell/droplet manipulation	[91]
Magnetic	Magnet	NdFeB	Sample	Dispenser	[92]
CNT-based	CNT-actuator	CNT	Sample	Dispenser	[93]
Microspheres	Heater	Polymer	Sample	Dispenser	[94]

Table 3. Lab-on-PCB devices and flow driving: a quantitative comparison.

Method	Module Dimension	Fabrication Complexity/Cost	Fluidic Condition	Channel Section	Reference
Pressurized chamber	3 × 2 cm^2	Medium/Low	Average speed 41 mm/s	0.3 × 1 mm ^{2}	[40]
Pressurized chamber	7.5 × 4.5 cm^2	Medium/Low	Average speed 16.25 mm/s	0.3 × 0.5 mm ^{2}	[48]
Pressurized chamber	6 × 1 cm^2	High/High	Average speed 271 mm/s	0.35 × 0.5 mm ^{2}	[47]
Pressurized chamber	6 × 5 cm^2	Low /Low	Average flow rate 1.4 µL/s	1 × 1.1 mm ^{2}	[49]
EWOD	8.9 × 8.9 cm^2 1024 electrodes	Low /Low	Max speed 100 mm/s	No channel	[53]
EWOD	~14 × 10 cm^2 6 electrodes	Low /Low	Max speed 3 mm/s	No channel	[56]
EWOD	N/A 64 electrodes	Low/Low	Velocity 6–13 mm/s	No channel	[61]
EWOD	12.78 × 8.55 cm^2 N/A electrodes	Medium/Low	N/A	No channel	[63,64]
EWOD	5 × 5 cm^2 576 electrodes	High/Low	N/A	No channel	[58]
Osmotic	~4.5 × 4.5 cm^2 6 pumps	Medium/Low	Flow rate 1 µL/min	0.2 × 0.1 mm ^{2}	[69,70]
Osmotic	7.62 × 2.54 cm^2 1 pump	High/Low	Max speed 800 µm/s	0.3 × 0.07 mm ^{2}	[71]
Osmotic	2 × 1 cm^2 1 pump	Low/High	Max speed 750 µm/s	N/A	[72]
Osmotic	15 × 15 cm^2 1 pump	High/Medium	Flow rate 8 µL/min	Nanopore Membrane	[73]
Peristaltic	~10 × 6 cm^2 1 pump	Low/Low	Average flow rate 500 µL/min	N/A	[88]
Peristaltic	~0.4 × 0.25 cm^2 1 pump	Low/Low	Max flow rate 1500 µL/min	N/A	[86,87]
Paraffin	5.5 × 4 cm^2	High/Low	Max flow rate 240 µL/min	N/A	[76]
Thermoneumatic	10 × 6 cm^2	High/Low	Moved volumen 60 µL	Min 0.3 × 1 mm ^{2}	[11]
Electrolytic	10 × 6 cm^2	High /Low	Max flow rate 0.8 µL/min	Min 0.3 × 1 mm ^{2}	[11]
Electrolytic	10 × 6.5 cm^2	Medium/High	Flow rate-backpressure 31.6 mL/min–547 kPa	1.5 × 2.5 mm	[78,79]
Electrolytic	2.5 × 1 cm^2	Low/Low	Max flow rate 135 µL/min	1 × 0.025 mm	[80]
Electrolytic	8 × 6 cm^2	High/Low	Max flow rate 0.1–1 µL/s	N/A	[81]
Electrolytic	N/A	Medium/Low	Max flow rate 100 µL/min	0.15 × 0.06 mm	[82]
SAW	11 × 11 cm	Low/Medium	Max speed 40 mm/s	0.15 × 0.06 mm	[91]
Magnetic	~4 × 3 cm	Low/Low	N/A	1 × 1 mm	[93]

Funding: This research received no external funding.

Institutional Review Board Statement: Not applicable.

Informed Consent Statement: Not applicable.

Data Availability Statement: Data is contained within the article.

Acknowledgments: I would like to thank Mars Tu for his invaluable support. In addition, I want to thank the Microsystems Group of the University of Seville for giving me the opportunity to be a part of the group.

Conflicts of Interest: The authors declare no conflict of interest.

Abbreviations

The following abbreviations are used in this manuscript:

MDPI	Multidisciplinary Digital Publishing Institute
DOAJ	Directory of open access journals
TLA	Three letter acronym
LD	linear dichroism

References

1. Moschou, D.; Tserepi, A. The lab-on-PCB approach: Tackling the μTAS commercial upscaling bottleneck. *Lab Chip* **2017**, *17*, 1388–1405. [CrossRef] [PubMed]
2. Zhao, W.; Tian, S.; Huang, L.; Liu, K.; Dong, L. The review of Lab-on-PCB for biomedical application. *Electrophoresis* **2020**, *41*, 1433–1445. [CrossRef] [PubMed]
3. Abgrall, P.; Gue, A. Lab-on-chip technologies: Making a microfluidic network and coupling it into a complete microsystem—A review. *J. Micromechan. Microeng.* **2007**, *17*, R15. [CrossRef]
4. Dittrich, P.S.; Manz, A. Lab-on-a-chip: Microfluidics in drug discovery. *Nat. Rev. Drug Disc.* **2006**, *5*, 210–218. [CrossRef]
5. Haeberle, S.; Zengerle, R. Microfluidic platforms for lab-on-a-chip applications. *Lab Chip* **2007**, *7*, 1094–1110. [CrossRef]
6. Merkel, T.; Graeber, M.; Pagel, L. A new technology for fluidic microsystems based on PCB technology. *Sens. Act. A Phys.* **1999**, *77*, 98–105. [CrossRef]
7. Nguyen, N.T.; Hejazian, M.; Ooi, C.H.; Kashaninejad, N. Recent advances and future perspectives on microfluidic liquid handling. *Micromachines* **2017**, *8*, 186. [CrossRef]
8. Terry, S.C.; Jerman, J.H.; Angell, J.B. A gas chromatographic air analyzer fabricated on a silicon wafer. *IEEE Trans. Electr. Dev.* **1979**, *26*, 1880–1886. [CrossRef]
9. Harrison, D.J.; Glavina, P.; Manz, A. Towards miniaturized electrophoresis and chemical analysis systems on silicon: An alternative to chemical sensors. *Sens. Act. B Chem.* **1993**, *10*, 107–116. [CrossRef]
10. Boyd-Moss, M.; Baratchi, S.; Di Venere, M.; Khoshmanesh, K. Self-contained microfluidic systems: A review. *Lab Chip* **2016**, *16*, 3177–3192. [CrossRef]
11. Liu, R.H.; Yang, J.; Lenigk, R.; Bonanno, J.; Grodzinski, P. Self-contained, fully integrated biochip for sample preparation, polymerase chain reaction amplification, and DNA microarray detection. *Anal. Chem.* **2004**, *76*, 1824–1831. [CrossRef] [PubMed]
12. Salado, G.F. Self-Contained Microfluidic Platform for General Purpose Lab-on-Chip Using pcb-mems Technology. Ph.D. Thesis, Universidad de Sevilla, Sevilla, Spain, 2017.
13. Chang, Y.J.; Hui, Y. Progress of Microfluidics Based on Printed Circuit Board and its Applications. *Chin. J. Anal. Chem.* **2019**, *47*, 965–975. [CrossRef]
14. Vasilakis, N.; Papadimitriou, K.; Evans, D.; Morgan, H.; Prodromakis, T. The lab-on-PCB framework for affordable, electronic-based point-of-care diagnostics: From design to manufacturing. In Proceedings of the 2016 IEEE Healthcare Innovation Point-Of-Care Technologies Conference (HI-POCT), IEEE, Cancun, Mexico, 9–11 November 2016; pp. 126–129.
15. ultimatepcb. 2021. Available online: https://www.ultimatepcb.com/services/pcb-fabrication (accessed on 29 January 2021).
16. Jobst, G.; Moser, I.; Svasek, P.; Varahram, M.; Trajanoski, Z.; Wach, P.; Kotanko, P.; Skrabal, F.; Urban, G. Mass producible miniaturized flow through a device with a biosensor array. *Sens. Act. B Chem.* **1997**, *43*, 121–125. [CrossRef]
17. Wego, A.; Pagel, L. A self-filling micropump based on PCB technology. *Sens. Act. A Phys.* **2001**, *88*, 220–226. [CrossRef]
18. Wego, A.; Glock, H.W.; Pagel, L.; Richter, S. Investigations on thermo-pneumatic volume actuators based on PCB technology. *Sens. Act. A Phys.* **2001**, *93*, 95–102. [CrossRef]
19. Wego, A.; Richter, S.; Pagel, L. Fluidic microsystems based on printed circuit board technology. *J. Micromech. Microeng.* **2001**, *11*, 528. [CrossRef]
20. Gaßmann, S.; Ibendorf, I.; Pagel, L. Realization of a flow injection analysis in PCB technology. *Sens. Act. A Phys.* **2007**, *133*, 231–235. [CrossRef]
21. Tsao, C.W.; DeVoe, D.L. Bonding of thermoplastic polymer microfluidics. *Microfluid. Nanofluidics* **2009**, *6*, 1–16. [CrossRef]
22. Thorsen, T.; Maerkl, S.J.; Quake, S.R. Microfluidic large-scale integration. *Science* **2002**, *298*, 580–584. [CrossRef]
23. Wang, Y.; Lin, W.Y.; Liu, K.; Lin, R.J.; Selke, M.; Kolb, H.C.; Zhang, N.; Zhao, X.Z.; Phelps, M.E.; Shen, C.K.; et al. An integrated microfluidic device for large-scale in situ click chemistry screening. *Lab Chip* **2009**, *9*, 2281–2285. [CrossRef]
24. Liu, J.; Jasim, I.; Shen, Z.; Zhao, L.; Dweik, M.; Zhang, S.; Almasri, M. A microfluidic based biosensor for rapid detection of Salmonella in food products. *PLoS ONE* **2019**, *14*, e0216873. [CrossRef]

25. Wilhelm, E.; Neumann, C.; Duttenhofer, T.; Pires, L.; Rapp, B.E. Connecting microfluidic chips using a chemically inert, reversible, multichannel chip-to-world-interface. *Lab Chip* **2013**, *13*, 4343–4351. [CrossRef]
26. Scott, A.; Au, A.K.; Vinckenbosch, E.; Folch, A. A microfluidic D-subminiature connector. *Lab Chip* **2013**, *13*, 2036–2039. [CrossRef] [PubMed]
27. Temiz, Y.; Lovchik, R.D.; Kaigala, G.V.; Delamarche, E. Lab-on-a-chip devices: How to close and plug the lab? *Microelectron. Eng.* **2015**, *132*, 156–175. [CrossRef]
28. Wagler, P.F.; Tangen, U.; Maeke, T.; Chemnitz, S.; Juenger, M.; McCaskill, J.S. Molecular systems on-chip (MSoC) steps forward for programmable biosystems. In *Smart Structures and Materials 2004: Smart Electronics, MEMS, BioMEMS, and Nanotechnology*; International Society for Optics and Photonics: San Diego, CA, USA 2004; Volume 5389, pp. 298–305.
29. Pittet, P.; Lu, G.N.; Galvan, J.M.; Ferrigno, R.; Blum, L.J.; Leca-Bouvier, B.D. PCB technology-based electrochemiluminescence microfluidic device for low-cost portable analytical systems. *IEEE Sens. J.* **2008**, *8*, 565–571. [CrossRef]
30. Kontakis, K.; Petropoulos, A.; Kaltsas, G.; Speliotis, T.; Gogolides, E. A novel microfluidic integration technology for PCB-based devices: Application to microflow sensing. *Microelectron. Eng.* **2009**, *86*, 1382–1384. [CrossRef]
31. Moschou, D.; Trantidou, T.; Regoutz, A.; Carta, D.; Morgan, H.; Prodromakis, T. Surface and electrical characterization of Ag/AgCl pseudo-reference electrodes manufactured with commercially available PCB technologies. *Sensors* **2015**, *15*, 18102–18113. [CrossRef]
32. Fu, Y.; Yuan, Q.; Guo, J. Lab-on-PCB-based micro-cytometer for circulating tumor cells detection and enumeration. *Microfluid. Nanofluidics* **2017**, *21*, 20. [CrossRef]
33. Sánchez, J.A.; Henry, O.; Joda, H.; Solnestam, B.W.; Kvastad, L.; Johansson, E.; Akan, P.; Lundeberg, J.; Lladach, N.; Ramakrishnan, D.; et al. Multiplex PCB-based electrochemical detection of cancer biomarkers using MLPA-barcode approach. *Biosens. Bioelectron.* **2016**, *82*, 224–232.
34. Cabello, M.; Aracil, C.; Perdigones, F.; Quero, J.M. Conditioning lab on PCB to control temperature and mix fluids at the microscale for biomedical applications. In Proceedings of the 2017 Spanish Conference on Electron Devices (CDE), IEEE, Barcelona, Spain, 8–10 February 2017; pp. 1–4.
35. Kaprou, G.D.; Papadopoulos, V.; Papageorgiou, D.P.; Kefala, I.; Papadakis, G.; Gizeli, E.; Chatzandroulis, S.; Kokkoris, G.; Tserepi, A. Ultrafast, low-power, PCB manufacturable, continuous-flow microdevice for DNA amplification. *Anal. Bioanal. Chem.* **2019**, *411*, 5297–5307. [CrossRef] [PubMed]
36. Cabello, M.; Mozo, M.; De la Cerda, B.; Aracil, C.; Diaz-Corrales, F.J.; Perdigones, F.; Valdes-Sanchez, L.; Relimpio, I.; Bhattacharya, S.S.; Quero, J.M. Electrostimulation in an autonomous culture lab-on-chip provides neuroprotection of a retinal explant from a retinitis pigmentosa mouse-model. *Sens. Act. B Chem.* **2019**, *288*, 337–346. [CrossRef]
37. Jolly, P.; Rainbow, J.; Regoutz, A.; Estrela, P.; Moschou, D. A PNA-based Lab-on-PCB diagnostic platform for rapid and high sensitivity DNA quantification. *Biosens. Bioelectron.* **2019**, *123*, 244–250. [CrossRef] [PubMed]
38. Dutta, G.; Regoutz, A.; Moschou, D. Enzyme-assisted glucose quantification for a painless Lab-on-PCB patch implementation. *Biosens. Bioelectron.* **2020**, *167*, 112484. [CrossRef] [PubMed]
39. El Fissi, L.; Fernández, R.; García, P.; Calero, M.; García, J.V.; Jiménez, Y.; Arnau, A.; Francis, L.A. OSTEMER polymer as a rapid packaging of electronics and microfluidic system on PCB. *Sens. Act. A Phys.* **2019**, *285*, 511–518. [CrossRef]
40. Flores, G.; Perdigones, F.; Aracil, C.; Quero, J. Pressurization method for controllable impulsion of liquids in microfluidic platforms. *Microelectron. Eng.* **2015**, *140*, 11–17. [CrossRef]
41. Moreno, J.; Perdigones, F.; Quero, J. Fabrication process of a SU-8 monolithic pressurized microchamber for pressure driven microfluidic applications. In Proceedings of the 8th Spanish Conference on Electron Devices, CDE'2011, IEEE, Palma de Mallorca, Spain, 8–11 October 2011; pp. 1–4.
42. Moreno, J.M.; Quero, J.M. A novel single-use SU-8 microvalve for pressure-driven microfluidic applications. *J. Micromech. Microeng.* **2009**, *20*, 015005. [CrossRef]
43. Aracil, C.; Quero, J.M.; Luque, A.; Moreno, J.M.; Perdigones, F. Pneumatic impulsion device for microfluidic systems. *Sens. Act. A Phys.* **2010**, *163*, 247–254. [CrossRef]
44. Perdigones, F.; Aracil, C.; Moreno, J.M.; Luque, A.; Quero, J.M. Highly integrable pressurized microvalve for portable SU-8 microfluidic platforms. *J. Microelectromech. Syst.* **2013**, *23*, 398–405. [CrossRef]
45. Aracil, C.; Perdigones, F.; Moreno, J.M.; Quero, J.M. BETTS: bonding, exposing and transferring technique in SU-8 for microsystems fabrication. *J. Micromech. Microeng.* **2010**, *20*, 035008. [CrossRef]
46. Flores, G.; Aracil, C.; Perdigones, F.; Quero, J. Low consumption single-use microvalve for microfluidic PCB-based platforms. *J. Micromech. Microeng.* **2014**, *24*, 065013. [CrossRef]
47. Aracil, C.; Perdigones, F.; Moreno, J.M.; Luque, A.; Quero, J.M. Portable Lab-on-PCB platform for autonomous micromixing. *Microelectron. Eng.* **2015**, *131*, 13–18. [CrossRef]
48. Flores, G.; Aracil, C.; Perdigones, F.; Quero, J.M. Lab-protocol-on-PCB: Prototype of a laboratory protocol on printed circuit board using MEMS technologies. *Microelectron. Eng.* **2018**, *200*, 26–31. [CrossRef]
49. Perdigones, F.; Franco, E.; Salvador, B.; Flores, G.; Quero, J.M. Highly integrable microfluidic impulsion system for precise displacement of liquids on lab on PCBs. *J. Microelectromech. Syst.* **2018**, *27*, 479–486. [CrossRef]
50. Cho, S.K.; Moon, H.; Kim, C.J. Creating, transporting, cutting, and merging liquid droplets by electrowetting-based actuation for digital microfluidic circuits. *J. Microelectromech. Syst.* **2003**, *12*, 70–80.

51. Choi, K.; Ng, A.H.; Fobel, R.; Wheeler, A.R. Digital microfluidics. *Ann. Rev. Anal. Chem.* **2012**, *5*, 413–440. [CrossRef]
52. Luo, J.; Fu, Y.Q.; Li, Y.; Du, X.; Flewitt, A.; Walton, A.; Milne, W. Moving-part-free microfluidic systems for lab-on-a-chip. *J. Micromech. Microeng.* **2009**, *19*, 054001. [CrossRef]
53. Umapathi, U.; Chin, S.; Shin, P.; Koutentakis, D.; Ishii, H. Scaling Electrowetting with Printed Circuit Boards for Large Area Droplet Manipulation. *MRS Adv.* **2018**, *3*, 1475–1483. [CrossRef]
54. Abdelgawad, M.; Wheeler, A.R. Rapid prototyping in copper substrates for digital microfluidics. *Adv. Mater.* **2007**, *19*, 133–137. [CrossRef]
55. Li, Y.; Chen, R.; Baker, R.J. A fast fabricating electro-wetting platform to implement large droplet manipulation. In Proceedings of the 2014 IEEE 57th International Midwest Symposium on Circuits and Systems (MWSCAS), IEEE, College Station, TX, USA, 3–6 August 2014; pp. 326–329.
56. Meimandi, A.; Seyedsadrkhani, N.; Jahanshahi, A. Development of an Electrowetting Digital Microfluidics Platform using Low-cost Materials. In Proceedings of the 2019 27th Iranian Conference on Electrical Engineering (ICEE), IEEE, Yazd, Iran, 30 April–2 May 2019; pp. 154–157.
57. Yi, Z.; Feng, H.; Zhou, X.; Shui, L. Design of an open electrowetting on dielectric device based on printed circuit board by using a parafilm m. *Front. Phys.* **2020**, *8*, 193. [CrossRef]
58. Nardecchia, M.; Lovecchio, N.; Llorca, P.R.; Caputo, D.; de Cesare, G.; Nascetti, A. 2-D digital microfluidic system for droplet handling using Printed Circuit Board technology. In Proceedings of the 2015 XVIII AISEM Annual Conference, IEEE, Trento, Italy, 3–5 February 2015; pp. 1–4.
59. Gong, M.; Kim, C.J. Two-dimensional digital microfluidic system by multilayer printed circuit board. In Proceedings of the 18th IEEE International Conference on Micro Electro Mechanical Systems, MEMS 2005, IEEE, Miami Beach, FL, USA, 30 January–3 February 2005; pp. 726–729.
60. Gong, J. All-electronic droplet generation on-chip with real-time feedback control for EWOD digital microfluidics. *Lab Chip* **2008**, *8*, 898–906. [CrossRef]
61. Gong, J.; Kim, C.J. Direct-referencing two-dimensional-array digital microfluidics using multilayer printed circuit board. *J. Microelectromech. Syst.* **2008**, *17*, 257–264. [CrossRef] [PubMed]
62. Braun, T.; Becker, K.F.; Koch, M.; Jung, E.; Lienemann, J.; Korvink, J.; Kahle, R.; Bauer, J.; Aschenbrenner, R.; Reichl, H. Contactless component handling on PCB using EWOD principles. In Proceedings of the 2008 10th Electronics Packaging Technology Conference, IEEE, Singapore, 9–12 December 2008; pp. 186–192.
63. Boles, D.J.; Benton, J.L.; Siew, G.J.; Levy, M.H.; Thwar, P.K.; Sandahl, M.A.; Rouse, J.L.; Perkins, L.C.; Sudarsan, A.P.; Jalili, R.; et al. Droplet-based pyrosequencing using digital microfluidics. *Anal. Chem.* **2011**, *83*, 8439–8447. [CrossRef] [PubMed]
64. Sista, R.; Hua, Z.; Thwar, P.; Sudarsan, A.; Srinivasan, V.; Eckhardt, A.; Pollack, M.; Pamula, V. Development of a digital microfluidic platform for point of care testing. *Lab Chip* **2008**, *8*, 2091–2104. [CrossRef]
65. Pollack, M.G.; Pamula, V.K.; Srinivasan, V.; Eckhardt, A.E. Applications of electrowetting-based digital microfluidics in clinical diagnostics. *Expert Rev. Mol. Diag.* **2011**, *11*, 393–407. [CrossRef] [PubMed]
66. Li, J. Current commercialization status of electrowetting-on-dielectric (EWOD) digital microfluidics. *Lab Chip* **2020**, *20*, 1705–1712. [CrossRef]
67. Wang, X.; Cheng, C.; Wang, S.; Liu, S. Electroosmotic pumps and their applications in microfluidic systems. *Microfluid. Nanofluidics* **2009**, *6*, 145–162. [CrossRef]
68. Li, L.; Wang, X.; Pu, Q.; Liu, S. Advancement of electroosmotic pump in microflow analysis: A review. *Anal. Chim. Acta* **2019**, *1060*, 1–16. [CrossRef]
69. Gassmann, S.; Pagel, L.; Luque, A.; Perdigones, F.; Aracil, C. Fabrication of electroosmotic micropump using PCB and SU-8. In Proceedings of the IECON 2012 38th Annual Conference on IEEE Industrial Electronics Society, IEEE, Montreal, QC, Canada, 25–28 October 2012; pp. 3958–3961.
70. Luque, A.; Soto, J.M.; Perdigones, F.; Aracil, C.; Quero, J.M. Electroosmotic impulsion device for integration in PCB-MEMS. In Proceedings of the 2013 Spanish Conference on Electron Devices, IEEE, Valladolid, Spain, 12–14 February 2013; pp. 119–122.
71. Wu, L.L.; Babikian, S.; Li, G.P.; Bachman, M. Microfluidic printed circuit boards. In Proceedings of the 2011 IEEE 61st electronic components and technology conference (ECTC), IEEE, Lake Buena Vista, FL, USA, 31 May–3 June 2011; pp. 1576–1581.
72. Mirzajani, H.; Cheng, C.; Wu, J.; Ivanoff, C.S.; Aghdam, E.N.; Ghavifekr, H.B. Design and characterization of a passive, disposable wireless AC-electro lab-on-a-film for particle and fluid manipulation. *Sens. Act. B Chem.* **2016**, *235*, 330–342. [CrossRef]
73. Babikian, S.; Jinsenji, M.; Bachman, M.; Li, G. Surface Mount Electroosmotic Pump for Integrated Microfluidic Printed Circuit Boards. In Proceedings of the 2018 IEEE 68th Electronic Components and Technology Conference (ECTC), IEEE, San Diego, CA, USA, 29 May–1 June 2018; pp. 498–503.
74. Ogden, S.; Klintberg, L.; Thornell, G.; Hjort, K.; Bodén, R. Review on miniaturized paraffin phase change actuators, valves, and pumps. *Microfluid. Nanofluidics* **2014**, *17*, 53–71. [CrossRef]
75. Walsh, D.; Zoller, P. *Standard Pressure Volume Temperature Data for Polymers*; CRC Press: Boca Raton, FL, USA, 1995.
76. Bodén, R.; Lehto, M.; Margell, J.; Hjort, K.; Schweitz, J.Å. On-chip liquid storage and dispensing for lab-on-a-chip applications. *J. Micromech. Microeng.* **2008**, *18*, 075036.
77. Wang, Y.; Li, Z.; Huang, X.; Ji, W.; Ning, X.; Liu, K.; Tan, J.; Yang, J.; Ho, H.p.; Wang, G. On-board control of wax valve on active centrifugal microfluidic chip and its application for plasmid DNA extraction. *Microfluid. Nanofluidics* **2019**, *23*, 112. [CrossRef]

78. Kim, H.; Hwang, H.; Baek, S.; Kim, D. Design, fabrication and performance evaluation of a printed-circuit-board microfluidic electrolytic pump for lab-on-a-chip devices. *Sens. Act. A Phys.* **2018**, *277*, 73–84. [CrossRef]
79. Kim, H.; Hwang, H.; Kim, J.; Kim, D. An Electrolytic Micropump Fabricated on Printed Circuit Board for Integrated Microfluidic System. In Proceedings of the 21st International Conference on Miniaturized Systems for Chemistry and Life Sciences, Savannah, GA, USA, 22–26 October 2017; pp. 641–642.
80. Pagonis, D.; Petropoulos, A.; Kaltsas, G. A pumping actuator implemented on a PCB substrate by employing water electrolysis. *Microelectron. Eng.* **2012**, *95*, 65–70. [CrossRef]
81. Schumacher, S.; Nestler, J.; Otto, T.; Wegener, M.; Ehrentreich-Förster, E.; Michel, D.; Wunderlich, K.; Palzer, S.; Sohn, K.; Weber, A.; et al. Highly-integrated lab-on-chip system for point-of-care multiparameter analysis. *Lab Chip* **2012**, *12*, 464–473. [CrossRef] [PubMed]
82. Li, J.; Wang, Y.; Dong, E.; Chen, H. USB-driven microfluidic chips on printed circuit boards. *Lab Chip* **2014**, *14*, 860–864. [CrossRef]
83. Liu, R.H.; Bonanno, J.; Yang, J.; Lenigk, R.; Grodzinski, P. Single-use, thermally actuated paraffin valves for microfluidic applications. *Sens. Act. B Chem.* **2004**, *98*, 328–336. [CrossRef]
84. Wang, Y.N.; Fu, L.M. Micropumps and biomedical applications—A review. *Microelectron. Eng.* **2018**, *195*, 121–138. [CrossRef]
85. Laser, D.J.; Santiago, J.G. A review of micropumps. *J. Micromech. Microeng.* **2004**, *14*, R35. [CrossRef]
86. Huang, X.; Nguyen, N.T. Development of a peristaltic pump in printed circuit boards. *J. Micromechatron.* **2005**, *3*, 1–13.
87. Nguyen, N.T.; Huang, X. Miniature valveless pumps based on printed circuit board technique. *Sens. Act. A Phys.* **2001**, *88*, 104–111. [CrossRef]
88. Wang, D.H.; Tang, L.K.; Peng, Y.H.; Yu, H.Q. Principle and structure of a printed circuit board process–based piezoelectric microfluidic pump integrated into printed circuit board. *J. Intel. Mater. Syst. Struc.* **2019**, *30*, 2595–2604. [CrossRef]
89. Schmid, L.; Wixforth, A.; Weitz, D.A.; Franke, T. Novel surface acoustic wave (SAW)-driven closed PDMS flow chamber. *Microfluid. Nanofluidics* **2012**, *12*, 229–235. [CrossRef]
90. Weser, R.; Darinskii, A.; Weihnacht, M.; Schmidt, H. Experimental and numerical investigations of mechanical displacements in surface acoustic wave bounded beams. *Ultrasonics* **2020**, *106*, 106077. [CrossRef] [PubMed]
91. Mikhaylov, R.; Wu, F.; Wang, H.; Clayton, A.; Sun, C.; Xie, Z.; Liang, D.; Dong, Y.; Yuan, F.; Moschou, D.; et al. Development and characterisation of acoustofluidic devices using detachable electrodes made from PCB. *Lab Chip* **2020**, *20*, 1807–1814. [CrossRef] [PubMed]
92. Hintermüller, M.A.; Jakoby, B. A Diffusor/Nozzle Pump Based on a Magnetically Actuated Flexible PCB Diaphragm. *Proceedings* **2018**, *2*, 1077.
93. Addinall, R.; Sugino, T.; Neuhaus, R.; Kosidlo, U.; Tonner, F.; Glanz, C.; Kolaric, I.; Bauernhansl, T.; Asaka, K. Integration of CNT-based actuators for bio-medical applications—Example printed circuit board CNT actuator pipette. In Proceedings of the 2014 IEEE/ASME International Conference on Advanced Intelligent Mechatronics, IEEE, Besacon, France, 8–11 July 2014; pp. 1436–1441.
94. Roxhed, N.; Rydholm, S.; Samel, B.; van der Wijngaart, W.; Griss, P.; Stemme, G. Low cost device for precise microliter range liquid dispensing. In Proceedings of the 17th IEEE International Conference on Micro Electro Mechanical Systems. Maastricht MEMS 2004 Technical Digest, IEEE, Maastricht, The Netherlands, 25–29 January 2004; pp. 326–329.
95. Narayanamurthy, V.; Jeroish, Z.; Bhuvaneshwari, K.; Bayat, P.; Premkumar, R.; Samsuri, F.; Yusoff, M.M. Advances in passively driven microfluidics and lab-on-chip devices: A comprehensive literature review and patent analysis. *RSC Adv.* **2020**, *10*, 11652–11680. [CrossRef]
96. Vasilakis, N.; Papadimitriou, K.I.; Morgan, H.; Prodromakis, T. High-performance PCB-based capillary pumps for affordable point-of-care diagnostics. *Microfluid. Nanofluidics* **2017**, *21*, 103. [CrossRef] [PubMed]
97. Vasilakis, N.; Moschou, D.; Carta, D.; Morgan, H.; Prodromakis, T. Long-lasting FR-4 surface hydrophilisation towards commercial PCB passive microfluidics. *Appl. Surf. Sci.* **2016**, *368*, 69–75. [CrossRef]
98. Schmitz, A.; Tranitz, M.; Wagner, S.; Hahn, R.; Hebling, C. Planar self-breathing fuel cells. *J. Power Sources* **2003**, *118*, 162–171. [CrossRef]
99. Schmitz, A.; Wagner, S.; Hahn, R.; Uzun, H.; Hebling, C. Stability of planar PEMFC in printed circuit board technology. *J. Power Sources* **2004**, *127*, 197–205. [CrossRef]
100. Griss, P.; Andersson, H.; Stemme, G. Expandable microspheres for the handling of liquids. *Lab Chip* **2002**, *2*, 117–120. [CrossRef] [PubMed]
101. Hong, C.C.; Murugesan, S.; Kim, S.; Beaucage, G.; Choi, J.W.; Ahn, C.H. A functional on-chip pressure generator using solid chemical propellant for disposable lab-on-a-chip. *Lab Chip* **2003**, *3*, 281–286. [CrossRef] [PubMed]
102. Ahn, C.H. Disposable polymer "smart" lab-on-a-chip for point-of-care testing (POCT) in clinical diagnostics. In Proceedings of the 13th International Conference on Solid-State Sensors, Actuators and Microsystems, 2005, TRANSDUCERS'05, IEEE, Digest of Technical Papers, Seoul, Korea, 5–9 June 2005; Volume 1, pp. 437–440.
103. Edwards, A.M.; Brown, C.V.; Newton, M.I.; McHale, G. Dielectrowetting: The past, present and future. *Cur. Opin. Col. Interface Sci.* **2018**, *36*, 28–36. [CrossRef]
104. Geng, H.; Feng, J.; Stabryla, L.M.; Cho, S.K. Dielectrowetting manipulation for digital microfluidics: Creating, transporting, splitting, and merging of droplets. *Lab Chip* **2017**, *17*, 1060–1068. [CrossRef]

105. Martín, A.; Kim, J.; Kurniawan, J.F.; Sempionatto, J.R.; Moreto, J.R.; Tang, G.; Campbell, A.S.; Shin, A.; Lee, M.Y.; Liu, X.; et al. Epidermal microfluidic electrochemical detection system: Enhanced sweat sampling and metabolite detection. *ACS Sen.* **2017**, *2*, 1860–1868. [CrossRef] [PubMed]
106. Kaprou, G.D.; Papadopoulos, V.; Loukas, C.M.; Kokkoris, G.; Tserepi, A. Towards PCB-Based Miniaturized Thermocyclers for DNA Amplification. *Micromachines* **2020**, *11*, 258. [CrossRef]

Article

A Comparative Analysis of Printed Circuit Boards with Surface-Mounted and Embedded Components under Natural and Forced Convection

Maksim Korobkov [1], Fedor Vasilyev [1,*] and Vladimir Mozharov [2]

1 Department of Digital Technologies and Information Systems, Moscow Aviation Institute (National Research University), 125993 Moscow, Russia; josef_turok@bk.ru
2 Kiratech srl, 37135 Verona, Italy; v.a.mozharov@gmail.com
* Correspondence: fedor@niit.ru

Abstract: This article is dedicated to the research of the physical reliability of electronic devices. It consists of a comparative thermal analysis of the cooling efficiency of a surface-mounted and an embedded component on a printed circuit board. A simulated finite element model of heat distribution over a printed circuit board with a surface component was constructed. An experiment confirmed the objectivity of the modeling results. The component's temperature was then analyzed depending on the installation method (surface and embedded) and the cooling method (natural and forced with varying airflow velocities). The results showed that the temperature of the embedded component was less than the temperature of the surface-mounted component under natural convection and, in most cases, under forced convection (with an airflow velocity of forced cooling under 16 $\frac{m}{s}$).

Keywords: PCB reliability; embedded components; finite element analysis; thermal analysis

Citation: Korobkov, M.; Vasilyev, F.; Mozharov, V. A Comparative Analysis of Printed Circuit Boards with Surface-Mounted and Embedded Components under Natural and Forced Convection. *Micromachines* **2022**, *13*, 634. https://doi.org/10.3390/mi13040634

Academic Editor: Francisco Perdigones

Received: 28 February 2022
Accepted: 14 April 2022
Published: 17 April 2022

1. Introduction

The embedding of electronic components into the base of a printed circuit board (PCB) is a promising direction for developing modern electronic system components for applications such as measuring, navigation [1–3], and electric power systems [4,5]. Furthermore, this technology potentially reduces the size of manufactured electronic devices, increases the density of interconnections, allows the application of new design solutions, improves device performance by placing components closer together and reduces parasitic inductances and capacitances [6,7]. In addition, the use of embedded components should increase the manufacturability of devices because the production of such PCBs combines the manufacturing and assembly processes.

However, the issue of ensuring the reliability of both electronic devices, in general, [8–18] and PCBs with embedded components, in particular [19,20], is very relevant. It is especially necessary to research the processes that led to breakage (failure physics), and the development of design rules based on these processes to ensure the reliability of electronic devices (design for reliability) [21]. For example, it has already been experimentally proven that, under thermal cycling [22] and drop test conditions [23], PCBs with embedded components are more reliable than PCBs with surface-mounted components.

Below (Figure 1) is the distribution between the physical factors that most often cause failures in electronic modules [24–26]. Temperature fluctuations and vibrations are the main causes of deformation in PCBs and the connections on it. This leads to a breakdown in the electrical contact, detachment of the component, delamination of the base material and the appearance of cracks in it, as well as the destruction of metalized holes. Humidity and dust cause contamination, leading to the destruction of conductors and elements. In addition, humidity induces electrochemical migration processes: it causes short circuits between conductors and leads to the formation of conductive anodic filaments (CAF).

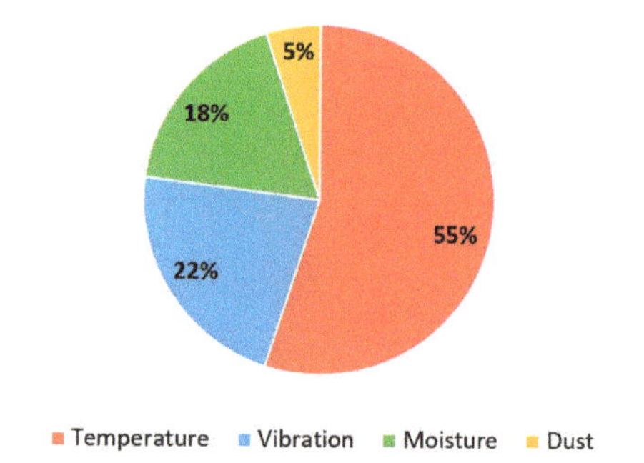

Figure 1. Distribution of the most frequent causes of electronic failures [22].

More than half of failures are due to thermal influences due to the wide variety of materials that make up the electronic device. They have different characteristics that should agree with each other. Below are some reasons for thermal failures in electronic modules:

1. the difference between the thermal expansion coefficients of the base material, electronic components, and solder leads to the destruction of solder joints;
2. the difference between the coefficients of thermal expansion of the base material and the material of the conductors leads to the destruction of metalized transition holes;
3. exposure to high temperatures, including heating of components in the absence of the required cooling system, can lead to PCB's deformations of various kinds, such as warping and twisting and its stratification.

Solving these problems is necessary, especially for power electronics. In addition, the use of embedded components can be used to increase temperature stability and improve the cooling efficiency of electronics. Embedding components inside PCBs allow, on the one hand, the provision of better heat transfer directly from the component but, on the other hand, also reduce the heat transferred by air cooling. Based on the above, a comparative analysis of thermal processes in PCBs with embedded components and PCBs with a traditional surface mounting method is of interest.

2. Materials and Methods

The experimental study of thermal processes is associated with some difficulties, including significant costs associated with the production of samples, the definition of test methods and the variation of parameters between experimental samples. Simulation modeling based on finite element analysis (FEA) facilitates such a study. However, very often, incorrect models are used.

Therefore, for thermal analysis the following were required:

1. The execution of an experiment to determine the heat distribution on the PCB with a surface component;
2. The construction of a simulation finite element model is constructed, the correctness of which is confirmed by the results of the experiment;
3. The expansion of a simulation model, and the execution of a comparative analysis of thermal processes in PCBs with embedded components and PCBs with a surface mounting method.

2.1. Experiment to Determine the Thermal Distribution on a PCB

The most straightforward heat source is a resistor attached to an FR-4 fiberglass work piece. By applying a voltage to the resistor terminals, we obtained a picture of the heat distribution on the work piece using a thermal imager.

A resistor was fixed with glue on a 1.5 mm thick FR-4 fiberglass billet in the 0805 case with 510 ± 1% Ohm resistance. When gluing, the maximum contact of the resistor with the fiberglass was ensured. Wires for connecting electric current were soldered to the side surfaces of the resistor. It should be noted that the wires soldered to the resistor were a strong heat sink. Thus, the diameter of the wires should be chosen as the minimum necessary. In our case, we wanted to ensure the heat dissipation of the resistor from 0.5 to 1 W, and we applied a voltage of 20 V to the resistor terminals. Then, the power allocated was 0.784 W, and the current was approximately 39 mA. Considering the short duration of exposure and the experimental nature of use, we used AWG 24 wires with a diameter of 0.51 mm to supply current. To further simplify the analysis of the heat distribution over the PCB, risks were applied to the board in 5 mm increments. The marks were needed to analyze the image obtained using a thermal imager. The created experimental model is shown in Figure 2.

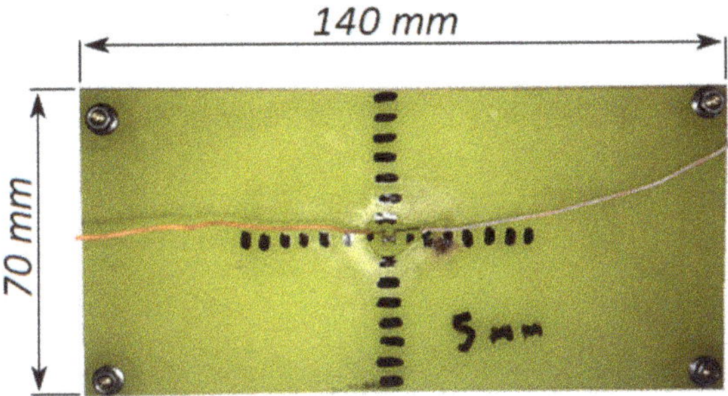

Figure 2. Experimental model.

The experiment involved heating the work piece in a vertical position. Below (Figure 3) is a picture of the heat distribution obtained during the experiment (the dissipated power was 0.784 W).

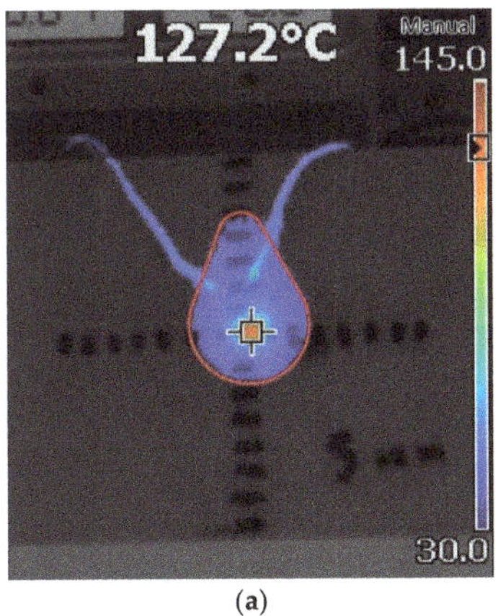

(**a**)

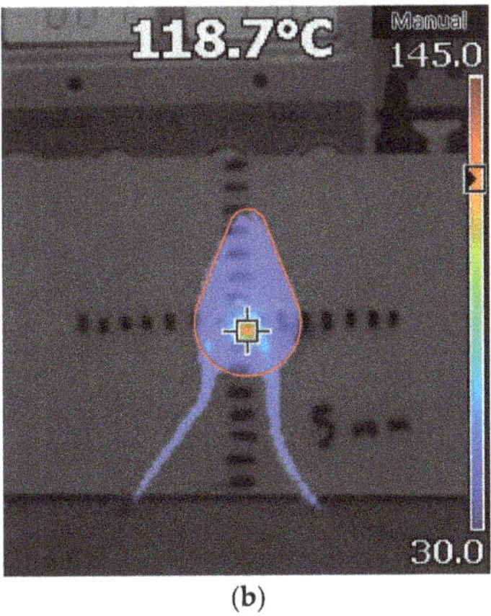

(**b**)

Figure 3. The picture of the heat distribution on the work piece during the experiment; the red contour highlights the heating area in which the temperature was above 303 K: (**a**) wires were laid over the work piece; (**b**) wires were laid under the work piece.

As can be seen, in the vertical position, a drop-shaped pattern was observed, which showed the fact of convection heating of the work piece over the component. According to the results of a series of measurements, the steady-state temperature was 405 ± 2 K. In addition, the arrangement of wires on the work piece did not affect the pattern of heat distribution: the results of the experiment with different arrangements of wires were almost identical (Figure 3).

Based on the obtained picture of the distribution of heat and temperature of the component, a future assessment of the correctness of the constructed simulation model was possible.

2.2. *Development of a Simulation Model of Thermal Processes*

To objectively model conduction and convection processes of heat transfer, we needed to choose software that allowed computational fluid dynamics (CFD) problems to be solved. Therefore, to solve this problem, the "Fluid Flow" module of the ANSYS software package was selected using the Fluent solver.

Also, several additional conditions needed to be imposed on the simulation model, taking into account all the particular features of the model, which are called boundary conditions [27]. The boundary conditions included the geometric model of the system, the physical properties of the system and the materials it consists of and the parameters of the process.

As a geometric model, a 3D model of the work piece was created, consisting of a $50 \times 50 \times 1.5$ mm FR-4 fiberglass base, with a simplified model of a $2 \times 1.2 \times 0.5$ mm resistor installed on it. Wires with a diameter of 0.51 mm and a length of 45 mm were connected to the resistor using solder. This length was because, in the experiment conducted at this length, the temperature that the wire was comparable to the ambient temperature. In addition, based on the experimental results, the work piece dimensions were reduced since this should not have affected the heat distribution pattern, but would increase the calculation speed.

To solve the CFD problems, it was necessary that the geometric model had a finite volume. To determine this volume, we placed the 3D model of the work piece in the center of the cylinder (Figure 4). The entire volume inside the cylinder, excluding the volume of the PCB, was filled with air. The cylinder was larger than the researched work piece to exclude the cylinder's influence on the simulation results: the outer diameter of the cylinder was 105 mm, the inner diameter was 100 mm, the height was 200 mm. Thus, a geometric model had been obtained, which needed to be divided into finite elements.

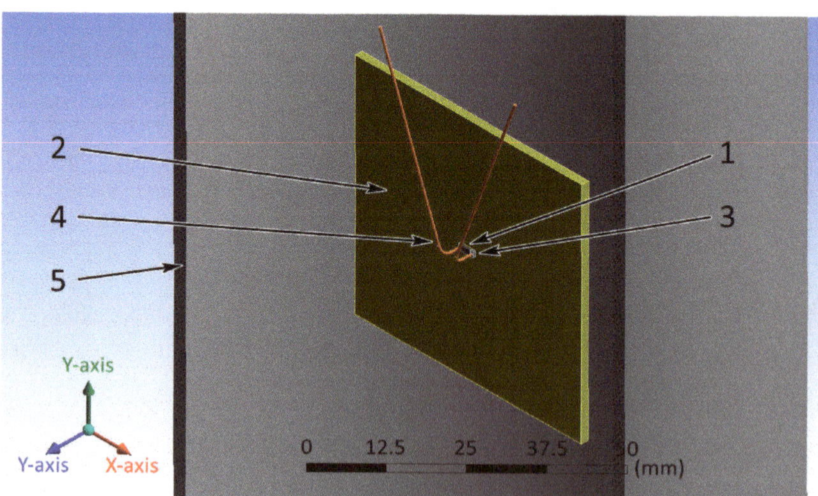

Figure 4. Cross-section of the final view of the geometric model: 1—component; 2—base; 3—solder; 4—wire; 5—cylinder.

In the model, the part of the airflow located next to the component and the base made of fiberglass was of the most significant interest since it carries out the convective heat transfer from the component. Therefore, the linear dimensions of the mesh in the component and base areas were reduced to 0.25 mm and 0.5 mm, respectively. As a result, the following grid was created on the geometric model (Figure 5). Below are the specified shape and the maximum allowable linear size of the mesh element for each model object (Table 1).

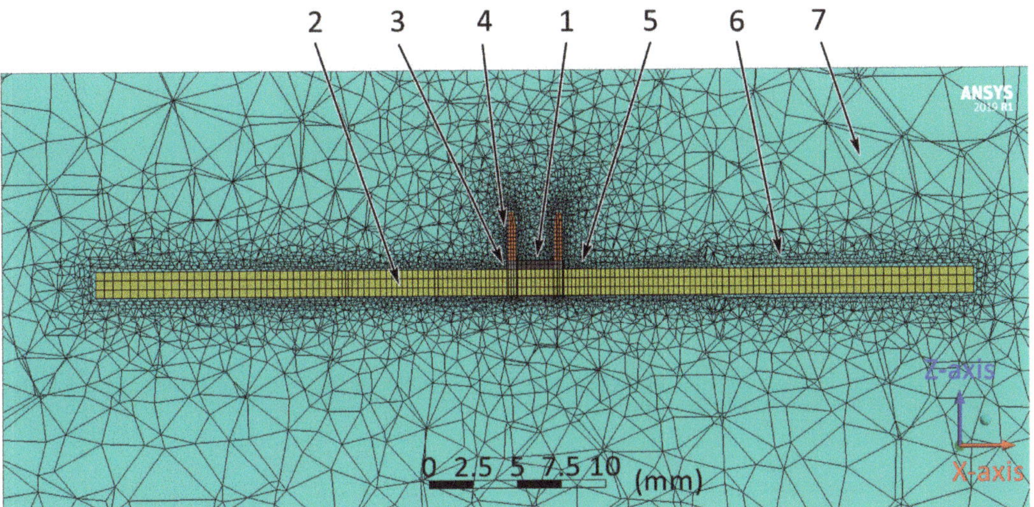

Figure 5. The cross-section of the grid created on the geometric model: 1—component; 2—base; 3—solder; 4—wire; 5—the air around the component, solder and wires; 6—the air around the base; 7—air.

Table 1. Mesh parameters.

Element of Model	Maximum Linear Size of Cell, mm	Shape of Cell
Component	0.25	Prism
Solder	0.25	Prism
Wires	0.25	Prism
Base	0.5	Prism
Cylinder	5	Prism
Air	5	Tetrahedron
Air (around component, solder and wires)	0.25	Tetrahedron
Air (around base)	0.5	Tetrahedron

The initial conditions of the process were also set:

1. The acceleration of gravity was opposite to the Y-axis (Figures 5 and 6) and was modulo 9.81 $\frac{m}{s^2}$;
2. The external pressure was 10^5 Pa;
3. The air density was set to a constant equal to 1.225 $\frac{kg}{m^3}$.

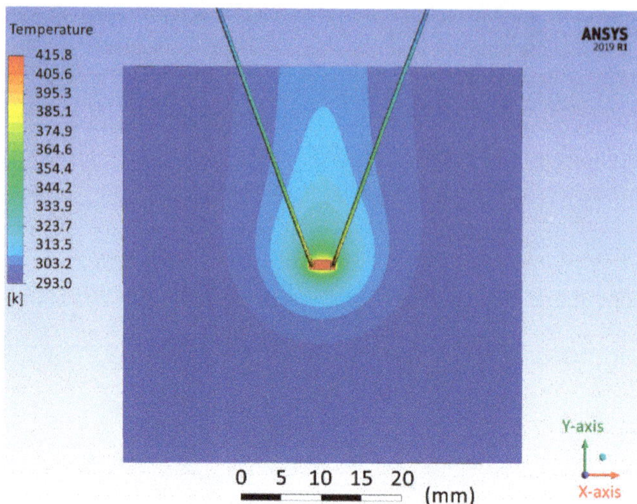

Figure 6. Heat distribution on the surface of the fiberglass base in the simulation model.

Next, the properties of solid materials were set, namely the following: density, specific heat capacity, thermal conductivity (Table 2).

Table 2. Properties of the solid materials used.

Element of Model	Material	Density, $\frac{kg}{m^3}$	Specific Heat Capacity, $\frac{J}{kg \cdot K}$	Thermal Conductivity, $\frac{W}{m \cdot K}$
Base	FR-4 fiberglass	1850	1300	0.343
Solder	Sn63Pb37	1890	250	82
Wires	Copper (Cu)	8900	381	395
Component	Alumina (Al_2O_3)	3690	880	18

It was also necessary to determine the air parameters (Table 3). Since the model did not assume the transition of air from a gaseous state to a liquid state, its behavior could be fully described by the model of an ideal gas.

Table 3. Air properties.

Specific Heat Capacity, $\frac{J}{kg \cdot K}$	1006.43
Thermal Conductivity, $\frac{W}{m \cdot K}$	$2.42 \cdot 10^{-2}$
Viscosity, $\frac{kg}{m \cdot s}$	$1.79 \cdot 10^{-5}$
Molecular Weight, $\frac{kg}{mol}$	$2.9 \cdot 10^{-2}$

Further clarification of the conditions of the process were made: during the construction, a standard type of pressure solver was used, as well as an absolute velocity model, since it was assumed that the airflow to the area under study occurred without vortices. The $k - \varepsilon$ model simulated the turbulence process [28]. The model was calculated using the SIMPLEC algorithm to calculate the arithmetic mean values of the scalar in adjacent cells separated by a face (green Gauss cell base).

2.3. *Analysis of the Constructed Simulation Model*

When comparing the obtained picture (Figure 6) with the result of the experiment (Figure 3), it could be seen that the results had a similar shape, with minor differences.

The component's temperature in the simulation model was 415.8 K, comparable to the temperature values obtained during the experiment (see Section 2.1).

Thus, the created model with sufficient accuracy reflected the investigated heating processes, which allowed it to be used in further experiments.

3. Results

The constructed simulation model gave us the opportunity to study the cooling processes of the PCB under conditions of natural and forced convection. In order to do this, we needed to set the direction of the forced airflow in the model. It could be assumed that the forced movement of air from top to bottom with small values of the airflow velocity would prevent the natural removal of heat from the PCB. However, how much this phenomenon affected the component's temperature was unknown. Therefore, to investigate this aspect in the model, the forced airflow was directed from top to bottom.

Four models were created based on the previously obtained results to analyze the effect of the airflow velocity on the temperature of the surface and embedded component (Figure 7).

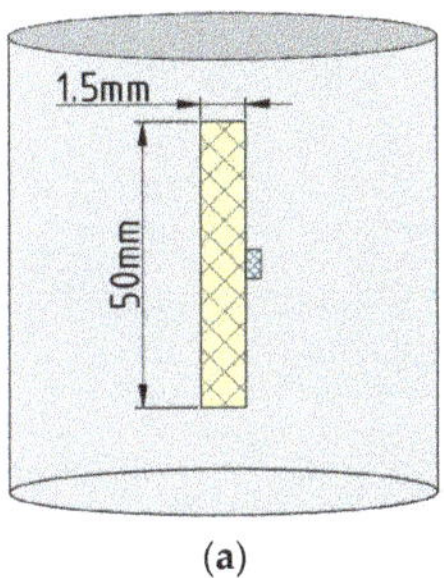

(a)

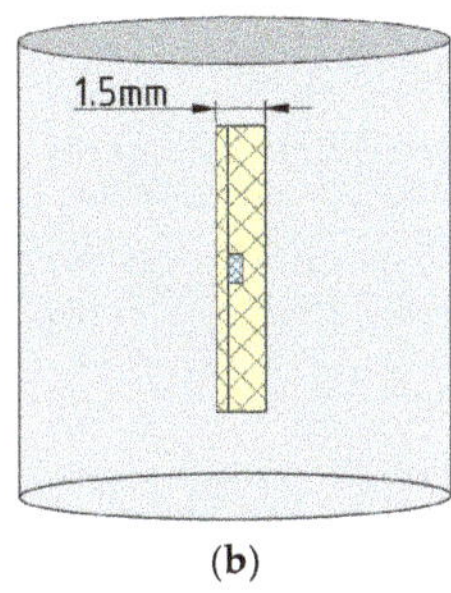

(b)

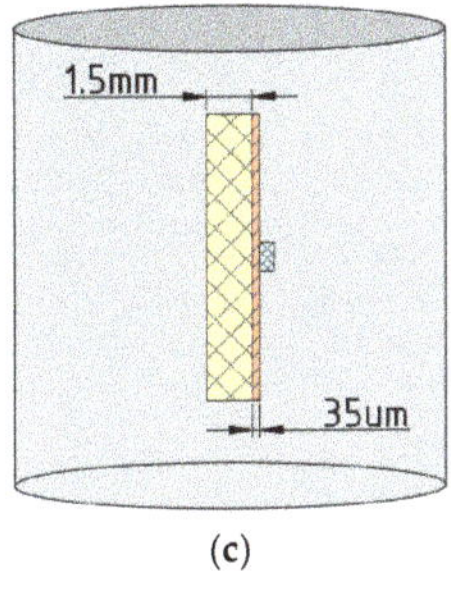

(c)

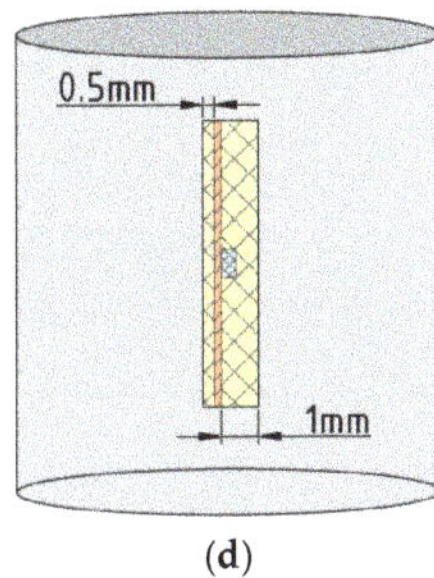

(d)

Figure 7. Geometric models of the experiment (side view): (**a**) surface component, without a copper layer; (**b**) embedded component, without a copper layer; (**c**) surface component, with a copper layer; (**d**) embedded component, with a copper layer.

Usually, in real printed circuit boards, the components have a permanent conductive heat sink through the conductors. In future modeling, we will consider only boundary cases. In the first case (Figure 7a,b), the components were attached directly to the fiberglass. In the second case (Figure 7c,d), the components were attached to a copper polygon without any conductive pattern.

Similar features of the models were a 50 × 50 × 1.5 mm FR-4 fiberglass base with an attached component 0805 was placed in the cylinder. The component, in turn, was a heat source with a volumetric power density equal to 0.654 $\frac{\text{W}}{\text{mm}^3}$. Again, the initial conditions described above were used (Sections 2.1 and 2.2).

Based on the simulation results (Figure 8), graphs of the component temperature dependence on the input current velocity (Figure 9) were constructed for models without a copper layer and the following conclusions were made:

1. Under conditions of natural and forced convection (at airflow values up to 16 $\frac{\text{m}}{\text{s}}$), the temperature of the embedded component was lower than the temperature of the surface component. This phenomenon was because the embedded component dissipated its power only into the base material, which increased the dispersion area and cooling efficiency. On the other hand, fiberglass has a low thermal conductivity, which did not allow effective cooling of the built-in component at high airflow rates.
2. The components' temperature increased at low airflow velocity values, confirming the assumption made earlier. However, it could be noticed that the temperature of the components reached its maximum at a forced flow rate of 0.25 $\frac{\text{m}}{\text{s}}$. This velocity value could be considered the absolute value of air velocity due to natural convection.

In this case, the component temperatures were 11 K and 9 K higher than the natural cooling for the surface and embedded components.

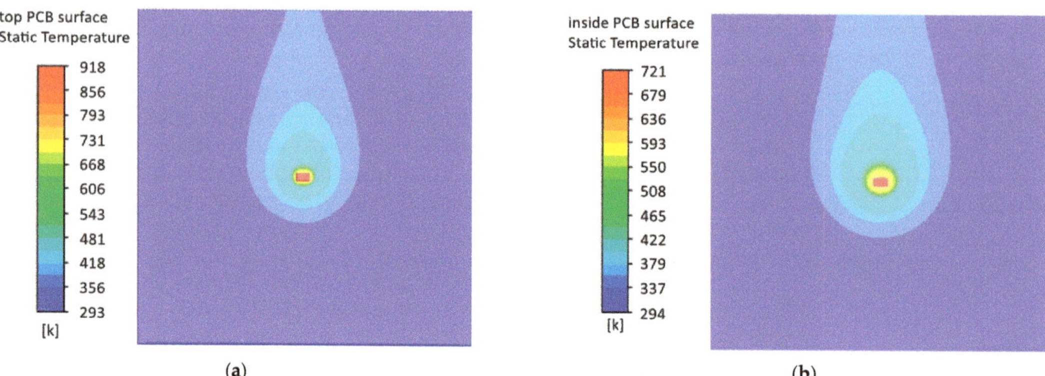

Figure 8. Heat distribution according to models without a copper layer under natural convection conditions: (**a**) surface component; (**b**) embedded component.

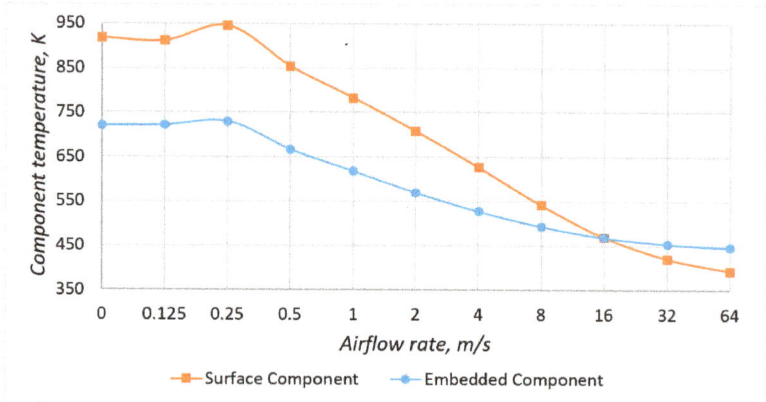

Figure 9. Dependence of the component temperature on the airflow rate in models without a copper layer.

From a practical point of view, the existing equipment cooling systems produced a flow rate of no more than 15 $\frac{m}{s}$. However, in real conditions, the airflow velocity in electronic equipment does not exceed 3 $\frac{m}{s}$. Therefore, the following conclusion could be drawn: in most use cases, namely in conditions of natural cooling and low-performance forced cooling, the built-in component will have a lower temperature compared to the surface one, which would positively affect its reliability and the reliability of the electronic device as a whole.

Let us consider models with a copper layer (Figure 10) and graphs of the component temperature dependence on the input flow rate (Figure 11). The introduction of a copper layer significantly reduced the temperature of the component and the temperature gradient on the PCB. When the flow velocity of the blown air was less than 16 $\frac{m}{s}$, the temperature of the embedded component was lower than the temperature of the surface component, which was similar to models without a copper layer. Under natural cooling conditions, the difference between the components' temperatures was about 3 K, and in the presence of forced cooling was about 1.5 K. However, the number of components on the real PCBs was tens and hundreds, and their power could be higher than in the considered model. In this case, the advantage of the embedded mounting would be more significant.

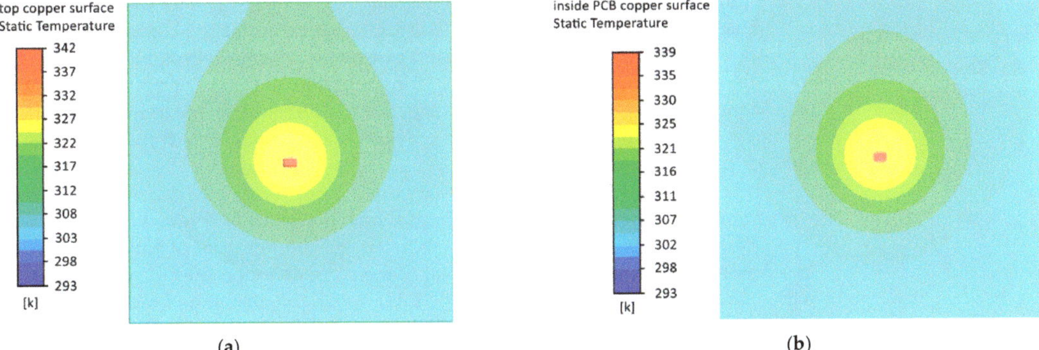

(**a**) (**b**)

Figure 10. Heat distribution according to models with a copper layer under natural convection conditions: (**a**) surface component; (**b**) embedded component.

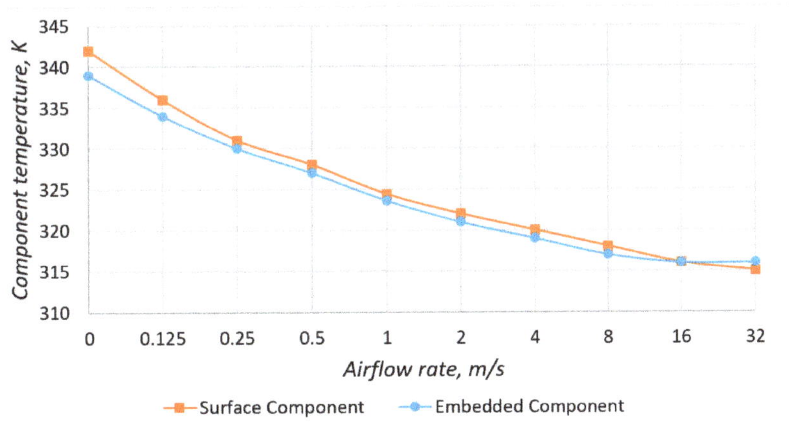

Figure 11. Dependence of the component temperature on the airflow rate in models with a copper layer.

Also, in models with a copper layer, there was no increase in temperature at a low airflow rate, which indicated that cooling naturally was insignificant in this case due to the slight difference between the temperatures of the PCB and the environment. Supplementary Materials for the article include the results of simulation of heat distribution by models under conditions of forced convection at different airflow rates.

4. Conclusions

Currently, the study of the physical reliability of electronic devices and the development of new methods to improve it are significant, especially for thermal processes. For these purposes, we created a simulation model of heat distribution on a printed circuit board, which took into account the processes of conductive and convective heat exchange. The reliability of the model was confirmed by experiment.

Then, based on the constructed model, we investigated the dependence of the component temperature on the cooling method (natural convection and forced convection with variation of the air flow). The results showed that the temperature of the embedded component was less than the temperature of the surface-mounted component under natural convection and, in most cases, of forced convection. The surface component cooled better under conditions of super-efficient forced convection, which are usually not available in real equipment. To do this, the airflow velocity needed to be greater than 16 $\frac{m}{s}$.

The simulation results showed that the temperature of the component, practically, did not depend on the installation method using a copper polygon. However, in modern multilayer printed circuit boards, polygons function as power supply routes. They are located in the middle of the PCB package, but not on its surfaces. In this case the surface component could be connected to the polygons only through the transition holes, which would further worsen its cooling.

It is important to note that the embedded component had the most significant advantage in the absence of forced cooling, which is typical for embedded systems. Thus, the embedded mounting technology reduced the temperature of the component. This phenomenon reduced the thermal effect of the component on the entire electronic device, which increased its overall reliability [29]. However, the question of the thermal efficiency of the integrated mounting with a high density of components on the inner layers of the printed circuit board remains open. In addition, in order to obtain a complete picture of the reliability of PCBs with embedded components, additional studies are necessary using conditions of vibration, humidity and dust.

Supplementary Materials: The following supporting information can be downloaded at: https://zenodo.org/record/6301748#.YhwXqehBzct (Accessed 13 April 2022).

Author Contributions: Conceptualization, F.V. and M.K.; methodology, F.V.; software, M.K.; validation, F.V. and V.M.; formal analysis, F.V.; investigation, M.K.; resources, F.V.; data curation, F.V. and V.M.; writing—original draft preparation, M.K.; writing—review and editing, V.M. and F.V.; visualization, M.K.; supervision, F.V.; project administration, F.V.; funding acquisition, F.V. All authors have read and agreed to the published version of the manuscript.

Funding: The work was performed within the framework of the state assignment of the Ministry of Education and Science (Russia), topic № FSFF-2020-0015.

Institutional Review Board Statement: Not applicable.

Data Availability Statement: Not applicable.

Conflicts of Interest: The authors declare no conflict of interest.

References

1. Busurin, V.; Korobkov, V.; Korobkov, K.; Koshevarova, N. Micro-Opto-Electro-Mechanical System Accelerometer Based on Coarse-Fine Processing of Fabry–Perot Interferometer Signals. *Meas. Tech.* **2021**, *63*, 883–890. [CrossRef]
2. Busurin, V.; Korobkov, K.; Shleenkin, L.; Makarenkova, N. Compensation Linear Acceleration Converter Based on Optical Tunneling. In Proceedings of the 2020 27th Saint Petersburg International Conference on Integrated Navigation Systems (ICINS), St. Petersburg, Russia, 25–27 May 2020. [CrossRef]
3. Busurin, V.; Zheglov, M.; Shleenkin, L.; Korobkov, K.; Bulychev, R. Development of an Algorithm to Suppress Frequency Splitting of an Axisymmetric Resonator of a Wave Solid-State Gyroscope with Optical Detection. *Meas. Tech.* **2020**, *62*, 879–884. [CrossRef]
4. Douine, B.; Berger, K.; Ivanov, N. Characterization of High-Temperature Superconductor Bulks for Electrical Machine Application. *Materials* **2021**, *14*, 1636. [CrossRef] [PubMed]
5. Kovalev, K.; Ivanov, N.; Zhuravlev, S.; Nekrasova, J.; Rusanov, D.; Kuznetsov, G. Development and testing of 10 kW fully HTS generator. *J. Phys. Conf. Ser.* **2021**, *1559*, 012137. [CrossRef]
6. Buttay, C.; Martin, C.; Morel, F.; Caillaud, R.; Le Leslé, J.; Mrad, R.; Degrenne, N.; Mollov, S. Application of the PCB-Embedding Technology in Power Electronics—State of the Art and Proposed Development. In Proceedings of the Second IEEE International Symposium on 3D Power Electronics Integration and Manufacturing, College Park, MD, USA, 25–27 June 2018.
7. Grübl, W.; Groß, S.; Schuch, B. Embedded Components for High Temperature Automotive Applications. In Proceedings of the 2018 IEEE 68th Electronic Components and Technology Conference, San Diego, CA, USA, 29 May–1 June 2018.
8. Vasilyev, F.; Isaev, V.; Korobkov, M. The Influence of the PCB Design and the Process of their Manufacturing on the Possibility of a Defect-Free Production. *Prz. Elektrotech.* **2021**, *97*, 91–96. [CrossRef]

9. Chvanova, M.; Vasilyev, F.; Isaev, V.; Baranov, V. Modeling Publication Terminology Maps on Quality Assessment Problems of Printed Circuit Boards. In Proceedings of the 2021 International Conference on Quality Management, Transport and Information Security, Information Technologies (IT&QM&IS), Yaroslavl, Russia, 6–10 September 2021. [CrossRef]
10. Vantsov, S.; Vasil'ev, F.; Medvedev, A.; Khomutskaya, O. Influence of Nonfunctional Contact Pads on Printed-Circuit Performance. *Russ. Eng. Res.* **2020**, *40*, 442–445. [CrossRef]
11. Vantsov, S.; Vasil'ev, F.; Medvedev, A.; Khomutskaya, O. Quasi-Determinate Model of Thermal Phenomena in Drilling Laminates. *Russ. Eng. Res.* **2018**, *38*, 1074–1076. [CrossRef]
12. Sokolsky, M.; Sokolsky, A. Electrochemical Migration: Stages and Prevention. *Amazon. Investig.* **2019**, *8*, 757–765.
13. Vasilyev, F.; Medvedev, A.; Barakovsky, F.; Korobkov, M. Development of the Digital Site for Chemical Processes in the Manufacturing of Printed Circuit Boards. *Inventions* **2021**, *6*, 48. [CrossRef]
14. Shashurin, V.; Vetrova, N.; Pchelintsev, K.; Kuimov, E.; Meshkov, S. Designing radio electronic systems for space purposes optimal by the criterion of reliability based on ultra-high-speed heterostructure nanoelectronics devices. *AIP Conf. Proc.* **2019**, *2171*, 150004. [CrossRef]
15. Meshkov, S.; Makeev, M.; Shashurin, V.; Tsvetkov, Y.; Khlopov, B. Microelectronics Devices Optimal Design Methodology with Regard to Technological and Operation Factors. In *International Scientific Conference Energy Management of Municipal Facilities and Sustainable Energy Technologies EMMFT 2018*; Murgul, V., Pasetti, M., Eds.; Advances in Intelligent Systems and Computing; Springer: Cham, Switzerland, 2018; Volume 982. [CrossRef]
16. Meshkov, S. Methodology of accounting technological and operational factors in the process of complex optimal design of micro and nanodevices, manufactured using group technologies. *J. Phys. Conf. Ser.* **2020**, *1560*, 012027. [CrossRef]
17. Makeev, M.; Meshkov, S. Study of degradation processes kinetics in ohmic contacts of resonant tunneling diodes based on nanoscale AlAs/GaAs heterostructures under influence of temperature. *AIP Conf. Proc.* **2017**, *1858*, 020001. [CrossRef]
18. Amosov, A.; Golikov, V.; Kapitonov, M.; Vasilyev, F.; Rozhdestvensky, O. Engineering and analytical method for estimating the parametric reliability of products by a low number of tests. *Inventions* **2022**, *7*, 24. [CrossRef]
19. Gorelov, A.; Vasilyev, F. 3D printed solder masks for printed circuit boards. *Period. Eng. Nat. Sci.* **2021**, *9*, 433–449. [CrossRef]
20. Schwerz, R.; Roellig, M.; Wolter, K.-J. Reliability analysis of encapsulated components in 3D-circuit board integration. In Proceedings of the 19th International Conference on Thermal, Mechanical and Multi-Physics Simulation and Experiments in Microelectronics and Microsystems (EuroSimE), Toulouse, France, 15–18 April 2018. [CrossRef]
21. Tulkoff, C.; Caswell, G. *Design for Excellence in Electronics Manufacturing*; Wiley: Hoboken, NJ, USA, 2021; p. 176.
22. Balmont, M.; Bord-Majek, I.; Ousten, Y. Comparative FEM thermo-mechanical simulations for built-in reliability: Surface mounted technology versus embedded technology for silicon dies. In Proceedings of the 21st European Microelectronics and Packaging Conference (EMPC) & Exhibition, Warsaw, Poland, 10–13 September 2017. [CrossRef]
23. Ryder, C. Embedded Components: A Comparative Analysis of Reliability. In Proceedings of the IPC APEX EXPO, Las Vegas, NV, USA, 12–14 April 2011.
24. Reynell, M. Advanced thermal analysis of packaged electronic systems using computational fluid dynamics techniques. *Comput.-Aided Eng. J.* **1990**, *7*, 104–106. [CrossRef]
25. Khayrnasov, K. Modeling and Thermal Analysis of Heat Sink Layers of Multilayer Board. *Amazon. Investig.* **2019**, *8*, 664–670.
26. Khayrnasov, K. Simulation of thermal conditions of a radio-electronic block of a cassette design. *Amazon. Investig.* **2019**, *8*, 671–677.
27. Denisov, M. *Mathematical Modeling of Thermophysical Processes. ANSYS and CAE-Design: Textbook*; UrFU: Yekaterinburg, Russia, 2011; p. 149.
28. Wilcox, D. *Turbulence Modelling for CFD*, 3rd ed.; DCW Industries: La Canada, CA, USA, 2006; p. 522.
29. Stiller, V. *Arrhenius Equation and Nonequilibrium Kinetics*; Mir: Moscow, Russia, 2000; p. 176.

 micromachines

Article

Semi-Automatic Lab-on-PCB System for Agarose Gel Preparation and Electrophoresis for Biomedical Applications

Jesús David Urbano-Gámez, Francisco Perdigones * and José Manuel Quero

Electronic Engineering Department, University of Seville, 41004 Sevilla, Spain; jurbanol@us.es (J.D.U.-G.); quero@us.es (J.M.Q.)
* Correspondence: fperdigones@us.es

Abstract: In this paper, a prototype of a semi-automatic lab-on-PCB for agarose gel preparation and electrophoresis is developed. The dimensions of the device are 38 × 34 mm^2 and it includes a conductivity sensor for detecting the TAE buffer (Tris-acetate-EDTA buffer), a microheater for increasing the solubility of the agarose, a negative temperature coefficient (NTC) thermistor for controlling the temperature, a light dependent resistor (LDR) sensor for measuring the transparency of the mixture, and two electrodes for performing the electrophoresis. The agarose preparation functions are governed by a microcontroller. The device requires a PMMA structure to define the wells of the agarose gel, and to release the electrodes from the agarose. The maximum voltage and current that the system requires are 40 V to perform the electrophoresis, and 1 A for activating the microheater. The chosen temperature for mixing is 80 °C, with a mixing time of 10 min. In addition, the curing time is about 30 min. This device is intended to be integrated as a part of a larger lab-on-PCB system for DNA amplification and detection. However, it can be used to migrate DNA amplified in conventional thermocyclers. Moreover, the device can be modified for preparing larger agarose gels and performing electrophoresis.

Keywords: lab-on-PCB; electrophoresis; biomedical applications; agarose

Citation: Urbano-Gámez, J.D.; Perdigones, F.; Quero, J.M. Semi-Automatic Lab-on-PCB System for Agarose Gel Preparation and Electrophoresis for Biomedical Applications. *Micromachines* **2021**, *12*, 1071. https://doi.org/10.3390/mi12091071

Academic Editors: Nam-Trung Nguyen and Angeliki Tserepi

Received: 5 August 2021
Accepted: 30 August 2021
Published: 2 September 2021

1. Introduction

Nowadays, the development of microfluidic devices using printed circuit board (PCB) substrates has been the subject of increasing research [1,2]. These devices are named lab-on-PCBs (LoP), and they can be considered to be a part of lab on a chip devices (LoCs). The use of PCB substrates has interesting advantages for biomedical applications [3], such as the possibility of easy integration of electronics and microfluidics with sensors and actuators in a single platform; commercial availability and low cost production, to name a few. It is important to emphasise that lab-on-PCBs can include several laboratory tasks, such as micromixing, sensing, chemical reactions and heating in a device with the dimensions of a credit card.

Apart from the PCB substrate, lab-on-PCBs was fabricated using different rapid prototyping materials, such as SU-8 [4] and PDMS [5,6]. However, the use of thermoplastics for the microfluidic component of the device is a more interesting option from the point of view of the market. In this respect, the device fabrication can be intended as mass production, that is, thermoplastic fabrication using hot embossing or injection molding, and the PCB can be ordered to specialised companies. These characteristics make lab-on-PCBs an attractive choice, due to their high potential for commercialisation [1,2].

Up to now, lab-on-PCBs include several biomedical applications, for example, for detecting cell viability [7], molecular diagnosis [8], organotypic cultures [9], or electrolytes detection [10]. Focusing on electrophoresis-based applications, capillary electrophoresis (CE) is the most usual technique in lab on chip [11]. This is because this kind of electrophoresis may be automated, and direct quantification is possible. Among others, CE is used for the separation of monosaccharides, oligosaccharides, and polysaccharides [12], separation of proteins [13] and for applications in life sciences in general [14]. On the other hand, conventional electrophoresis continues being a robust and very used method [15] both for lab on a chip applications [16,17] and conventional laboratories [18].

Whatever the method, the electrophoretic gel (agarose or polyacrylamide) is manually fabricated in many laboratories. These methods require heating, time to achieve transparency and a curing time to solidify the material. The current methods suggested for many agarose manufacturers can be seen in [19–21], to name a few. As previously commented, the procedure is manual. In this respect, Erlenmeyer flasks or beakers are required. When the agarose is placed in the buffer, it is insoluble at room temperature. However, when the agarose solution is heated, the agarose particles increase the solubility and they become hydrated, and therefore go into solution. This heating is performed using a microwave at high power. Moreover, the suggested method to stop the mixing process is boiling and transparency. However, the transparency is subjective because it depends on the expertise of a technician, without an exact measure of the transparency. Furthermore, the process implies big equipment when compared to the functionalities that a lab on a chip can offer. On the other hand, there are companies, for instance Invitrogen^{TM} (Thermofisher Scientific), that supply the ready-to-use gel cartridge. In this case, the dimensions are fixed, they are not integrable on a lab on chip, nor are they customizable, and they are intended to be used in its reader. Moreover, the quantity of agarose for laboratories is limited to the commercial electrophoretic tanks dimensions. In this respect, new techniques are used for preparing the agarose gel [22]. The agarose gel electrophoresis is typically used to resolve RNA and DNA; polyacrylamide gel electrophoresis is used to separate proteins. Therefore, there is a wide variety of applications that are processed in laboratories.

The trend of the lab on chip developments includes the integration of these applications in a substrate with the dimensions of a credit card. In order to do so for qualitative PCR, the thermocycling, the agarose gel preparation, the electrophoresis and the detection have to be integrated in the same platform, with a process as automatic as possible. Regarding DNA amplification, many approaches have been reported [23–25]. Lab-on-PCB devices have also been developed, for example, for a three-temperature protocol [17] and for two temperatures [16]. All of them require the agarose gel preparation, using the typical procedure. Moreover, none of them have the preparation of the agarose integrated in the same platform. Regarding the electrophoresis on chip, there are many devices apart from CE [26,27], especially for polyacryamide gel. In these cases, photopatterning of polyacrylamide gels in glass or PDMS microfluidic devices is performed, in order to prepare the gel in the microchannels [28–33]. These devices demonstrate interesting biological applications, using materials fabricated with no low-cost processes. The device reported on [34] is used for analysing single-cell genomic damage. It is an interesting device fabricated using soft lithography for agarose gel and a SU-8 mold. The fabrication process of this device is manual, using rapid prototyping materials. In this case, the integration of the conventional electrophoretic method for DNA migration on lab on chip is challenging due to the low automation of the method. Finally, the detection method in an important component of the whole system. In this respect, inexpensive and single-use lab on a chip devices are not intended to include the detection system. These systems are composed of expensive components, especially for absorbance or fluorescence, for example, photodiodes, photomultipliers or phototransistors. These single-use lab on chips require a reader to perform the detection.

In this paper, a semi-automatic and disposable lab-on-PCB for preparing agarose gel and for performing electrophoresis is described. It includes a microheater to achieve the mixing by increasing the solubility of the agarose in the TAE buffer (Tris-acetate-EDTA buffer). Regarding the buffer, TBE buffer (Tris-borate-EDTA buffer) could be used as well. The device has integrated sensors, such as a thermistor to control the temperature and light-dependent resistor to ensure the required transparency. In addition, it includes an interdigitated capacitive sensor to detect the filling of the cavity with the liquid. All these sensors and actuators are connected to a microcontroller, which governs the working of the lab-on-PCB. The device testing includes preparing the agarose gel using the device, and performing the electrophoresis. The device implies low-cost and single-use characteristics, fast analysis, low sample consumption and integration capability. It is intended as an integrable functional module of a more complex system for DNA amplification and detection, or even a device itself to perform more controllable electrophoresis in conventional laboratories, minimizing the human factor.

2. Lab-on-PCB Brief Description

In this section, the function of the lab-on-PCB is briefly described. Then, in Section 3, the sensors and actuators are described for a complete understanding of the device.

The PCB substrate has two functional components; Figure 1. The first one is intended to control the agarose mixing and curing, and the second one to perform the electrophoresis. The dimensions of the PCB are 38 × 34 mm^2.

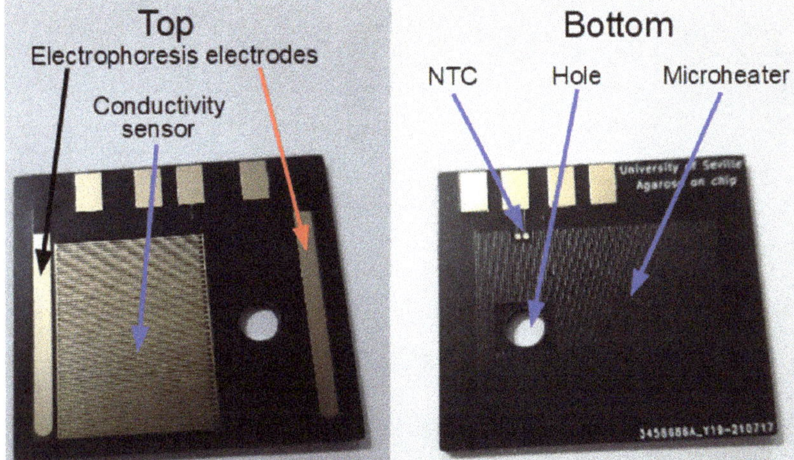

Figure 1. Printed circuit board substrate for agarose mixing and curing, and electrophoresis. (**Left**) Conductivity sensor and electrophoresis electrodes are shown. (**Right**) The microheater and the thermistor can be seen. The dimensions of the device are 38 × 34 mm^2.

Regarding the preparation of the agarose, the device includes a microheater in order to mix the agarose with the TAE buffer. Typically, the mixing is considered finished when the agarose is transparent enough. Therefore, the degree of transparency needs to be monitored. In order to do so, a light dependent resistor (LDR) sensor is used. In addition, the mixing temperature is measured with an integrated surface mounted device (SMD) thermistor to perform the control of temperature. On the other hand, the device includes a conductivity sensor in order to detect the filling of the cavity with the liquid to start the automatic process. These steps, that is, the mixing and the curing of the agarose are governed by a microcontroller. Finally, the electrophoresis is performed, using two gold electrodes integrated in the same PCB substrate.

The device has polymethylmethacrylate (PMMA) walls, which limit the area of the fabricated agarose gel. This part has an auxiliary structure to define both the wells in the agarose gel and the volume above the electrodes; Figure 2. The auxiliary structure has to be inserted in the cavity before the filling with the TAE buffer.

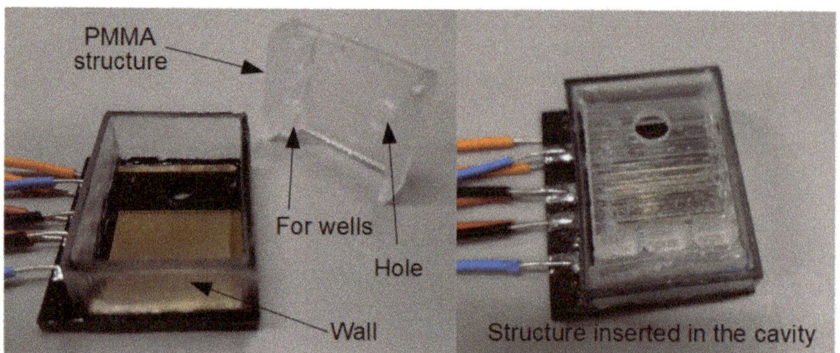

Figure 2. The lab-on-PCB with the thermoplastic wall and the auxiliary structure are shown.

In order to clarify the assembly, a drawing of a cross-sectional view of the lab-on-PCB is shown in Figure 3. As can be seen, the transparent film is placed on the top side of the PCB substrate, and the LDR sensor is located below the transparent film. The detection system is not included because it is an independent part of the system.

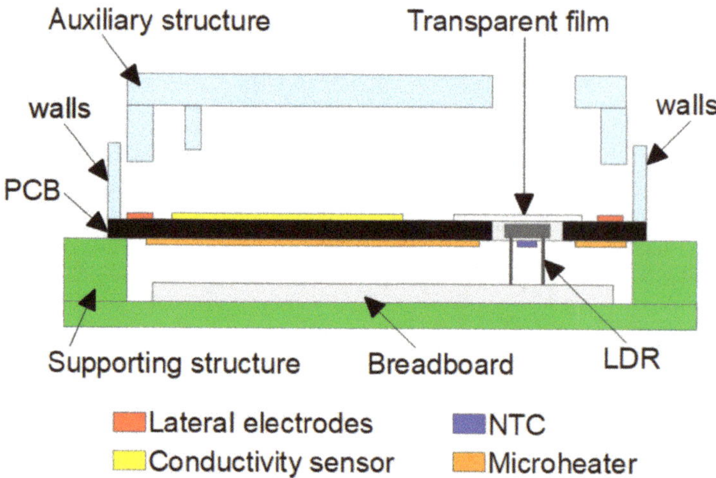

Figure 3. The cross-sectional view of the prototype is shown. The supporting structure, the breadboard and the location of the sensors and the transparent film can be seen. The negative temperature coefficient (NTC) sensor is not under the light dependent resistor (LDR), it is in a different plane.

The device requires a basic signal conditioning circuit in order to manage both the current along the microheater and the measure of the sensors outputs. The whole behaviour of the lab-on-PCB for agarose gel preparation is governed by a microcontroller. The schematic of the electronic circuit can be seen in Figure 4.

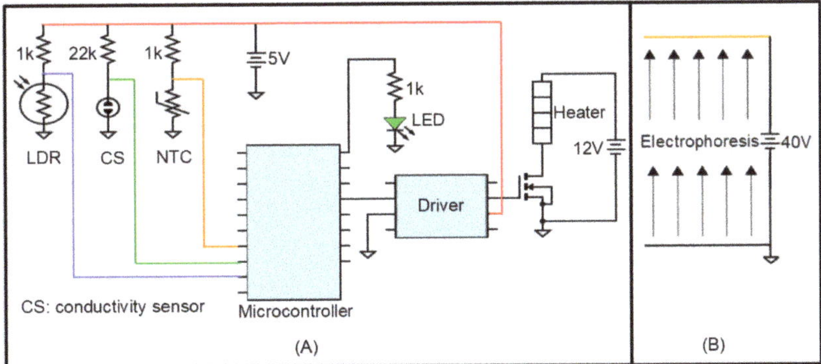

Figure 4. (**A**) Basic signal conditioning electronic circuit connected to the microcontroller. (**B**) The electrophoresis schematic circuit. The arrows indicate the direction of the migration.

3. Sensors and Actuators Description

3.1. Thermal Actuation

The microheater is directly integrated in the bottom side of the PCB substrate during fabrication, and covered with the solder mask. It is performed by the commercial PCB manufacturer with a very good finish. This microheater is a copper serpentine with a length of 2.1 m, a width of 150 µm, a spacing between lines of 150 µm, and a thickness of 35 µm. The experimental resistance of the copper microheater is 7 Ω. The control of the temperature is performed using a SMD thermistor (part number: NXFT15XH103. Resistance = 10 kΩ, B-constant = 3380 K and package 0402, Murata Manufacturing Co., Ltd., Kyoto, Japan) which is placed close to the microheater. The temperature needs to be characterised previously because the temperature of the liquid over the microheater and the thermistor temperature is not necessarily the same. This characterisation is described in Section 5.

3.2. Optical Sensing

The PCB substrate has a hole with a diameter of 4.5 mm, covered with a transparent and PCR-compatible film (ThermalSeal®films, classic). The optical sensor is a LDR (NSL-19M51) with a diameter of 4.3 mm and a minimum light resistance of 20 kΩ. It is placed just below the transparent film of the PCB hole. As can be seen in Figure 2, the auxiliary structure has a hole to avoid the lack of transparency due to the CNC laser fabrication process. This hole is just above the PCB hole in order to create an optical path, which allows to measure the transparency. The values of the optical sensor have to be characterised to define the transparency at the end of both the mixing and the curing. This characterisation is described in Section 5.

3.3. Conductivity Sensing

Similarly to the microheater, the conductivity sensor is fabricated by the manufacturer company, in this case, in the top side of the substrate and covered with gold. This sensor is an integrated interdigitated transducer (IDT) with 74 electrodes with a width of 150 µm, a gap of 150 µm, and a length of 1.8 cm. The dimensions of the electrodes and the microheater are limited by the selected technology, but they can be reduced by increasing the fabrication cost. This sensor requires characterisation in order to define the value for detecting the TAE buffer. This characterisation is described in Section 5 together with the optical and thermal experiments.

3.4. Electrophoretic Actuation

The migration of the DNA is performed by actuation on the lateral electrodes. These electrodes are integrated during the fabrication process. They are copper electrodes covered with gold for reducing the oxidation during the electrophoresis process. The electrodes have a length of 2.8 cm, a width of 2 mm and a thickness of 35 µm.

4. Process Sequence

The steps of the process can be seen in Table 1 and are commented thereafter.

Table 1. Whole process sequence for agarose gel preparation and electrophoresis.

	Agarose preparation	
Step	Function	Automatic
1	Filling of the cavity	First step
2	Mixing	Yes
3	Curing detection	Yes
4	Removing of the auxiliary structure	No
	Electrophoresis	
Step	Function	Automatic
5	DNA loading	No
6	Electrophoresis	No

The device with the PMMA structure assembled can include the agarose powder (CSL-AG500 Cleaver Scientific, Rugby, Warwickshire, UK) over the surface before starting the process, in this case 100 mg for a 4 ml agarose gel (final concentration 2.5% w/v). The first step consists of filling the cavity with the agarose-TAE mixture with SYBRSafe DNA staining solution (S33102 ThermoFisher Scientific, Waltham, Massachusetts, USA) to perform the mixing, where the TAE buffer is (15558042 ThermoFisher Scientific). This filling is performed using a syringe pump (NewEra Pump Systems NE-1000), with a volume of 4 mL. The percentage of agarose can be modified by changing the TAE buffer volume, the quantity of agarose or both of them. This step is detected by the conductivity sensor, which sends the signal for starting the automatic process. The next step consists of disabling the conductivity sensor, supplying the required current to the microheater, and sensing the degree of transparency. Once the transparency is achieved, the third step takes place automatically, that is, the microheater is disabled in order to cool down the mixture. In this step, the LDR sensor continues to be enabled because the degree of transparency of the cured agarose is lower than the freshly mixed agarose. The automatic process finishes when the agarose is cured, after which the microcontroller activates a LED in order to inform that the process is finished, and disable the sensors. Finally, the following step is not automatic; it consists of removing the PMMA structure to define both the wells and the cavities to pour the TAE buffer. After this process, the device is ready to be loaded with the liquid to be migrated, using electrophoresis. The experimental results show the performance of the fabricated agarose gel for electrophoresis.

Once the agarose gel is cured and the PMMA structure is removed from the substrate, the next step consists of loading the wells with DNA. After that, the electrophoresis is performed, using an independent power supply of 40 V (the rest of the power supplies are switched off), for which the lateral gold electrodes are used. In this case, these two steps are manually performed.

It is worth highlighting that the filling process can be performed manually, or even integrated in the automatic process. The last option would imply that the conductivity sensor would not be necessary for detecting the liquid. In addition, the mixing temperature and the required degree of transparency could be modified, if necessary. Finally, although the electrophoresis process is disconnected from the agarose fabrication, they can be joined by software. In order to do, automatic pipettes, and the integration of the actuation of the lateral electrodes are necessary. This is possible, but the low-cost nature of the device is lost. However, it could be interesting for large laboratories.

5. Results and Discussion

Before performing the experiment to prepare the agarose gel and define a program to control the process, the characterisation of the sensors and actuators is required.

The microheater characterisation consists of relating the temperature of the agarose-TAE solution with the temperature of the negative temperature coefficient (NTC) thermistor. In order to do so, a thermocouple is used for measuring the temperature of the liquid of the cavity (agarose-TAE solution). In addition, the current supplied to microheater has to be defined. The results for the microheater can be seen in Figure 5.

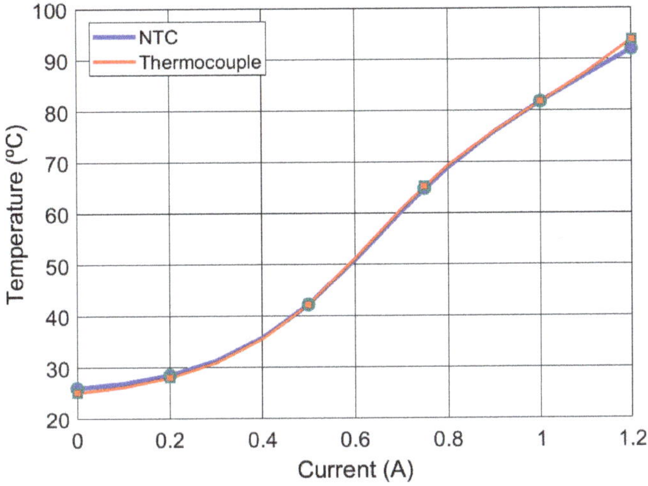

Figure 5. Temperature of both the agarose-TAE solution (thermocouple) and the NTC thermistor as a function of the current.

The temperature for performing the agarose gel is chosen to be a set point of 80 °C. This implies a thermistor resistance of 1680 Ω. The circuit to control the temperature is a voltage divider, so the voltage of the thermistor is applied to an analogue pin of the microcontroller (Figure 4). In addition, a driver (TC4427CPA) and a MOSFET (nMOS IRFB4227PBF) are required to supply a controlled current to the microheater.

The temperatures on the two sides of the PCB substrate are quite different. The surface at low temperature is in contact with the agarose-TAE solution, and the opposite side is in contact with air. Similarly to this effect, the temperature in the centre of the microheater is higher than the temperature where the NTC sensor is placed, both of them being on the bottom side of the PCB. These effects compensate for each other, and the temperatures of the NTC and the liquid (thermocouple) are very close, as can be seen in Figure 5. The authors have considered that the temperatures are the same for this device; however, for a different design of the device, this assumption could be incorrect. Three tests were performed to obtain the behaviour depicted in Figure 5. All of them showed a very similar behaviour, with the electrical current ranging between 0.95 and 1 A for a temperature of 80 °C. After that, the final experiments (automatic working) showed the same behaviour. Regarding the boundary conditions, the thermocouple was placed inside the liquid at approximately half its height. In addition, the room temperature was 25–26 °C. Measures to avoid convection were not used, and the conditions of the room imply natural convection.

The characterisation of the interdigitated conductivity sensor is simple to perform. The objective is to obtain a value of the conductivity to define a starting point of the process. In this case, the conductance before filling the cavity with the TAE buffer is 0 S. When the liquid is in the cavity, the conductance rapidly decreases up to 0.04 mS, which corresponds to a resistance of 20 kΩ. This value is chosen to define the starting point of the process; that is, the process starts when the conductivity sensor reaches 0.04 mS. The electronic circuit is a voltage divider. Finally, this characterisation is carried out using an oscilloscope (Tektronix TDS 2012B, single seq. "falling" procedure); Figure 6.

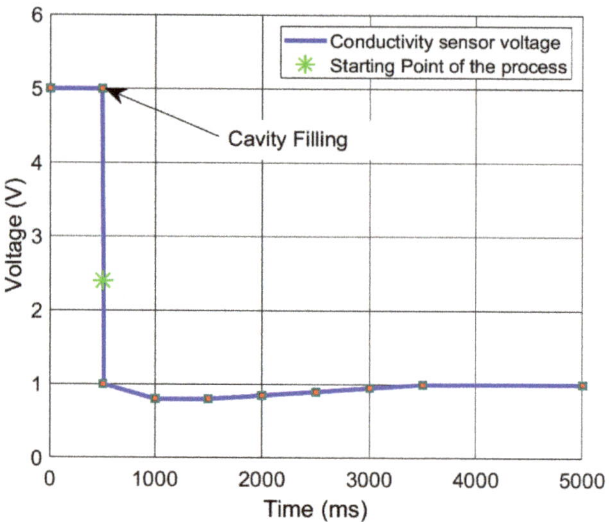

Figure 6. The output of the voltage divider as a function of the time is shown. The starting point is marked using an asterisk.

The characterisation of the optical sensor consists of measuring the degree of transparency of the agarose-TAE solution; Figure 7. In order to do so, another voltage divider is used, where the voltage of the LDR is measured. In this case, the agarose gel performed is 2.5% w/v in the TAE buffer. The choice of the final transparency of the mixture is defined, taking into account the expertise of the authors.

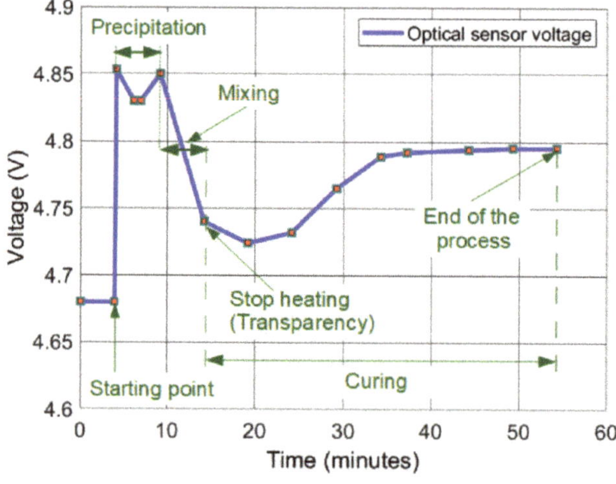

Figure 7. Different values of the optical sensor voltage as a function of the time.

The initial value of the optical sensor is 4.68 V before pouring the TAE buffer. After that, this value increases up to 4.85 V when the powder goes to the surface of the liquid. Then, the powder starts to precipitate and a slight transparency is achieved. When the powder finishes precipitating, that transparency is lost. At this point, the mixing starts, and the liquid begins to become transparent. Therefore, the voltage decreases. The final transparency is reached at 4.74 V, and the microheater is disabled. Later on, the liquid continues increasing the transparency during the cooling down. Then, the agarose starts curing. Once the transparency is lost again, the process finishes at V = 4.795 V. All these values have to be taken into account for programming the microcontroller (ATMega380P).

The process for preparing the agarose gel takes 40 min from the beginning, where the first seconds are used for filling with the TAE buffer, and the last seconds are for removing the PMMA structure. The rest of the process is automatic, where the mixing time is 10 min and the curing time is about 30 min.

Once the sensors and actuators are characterised and the microcontroller is programmed, the experiments are carried out. The resulting agarose gel after both the mixing and the curing is shown in Figure 8. In addition, the wells loaded with liquids can be seen.

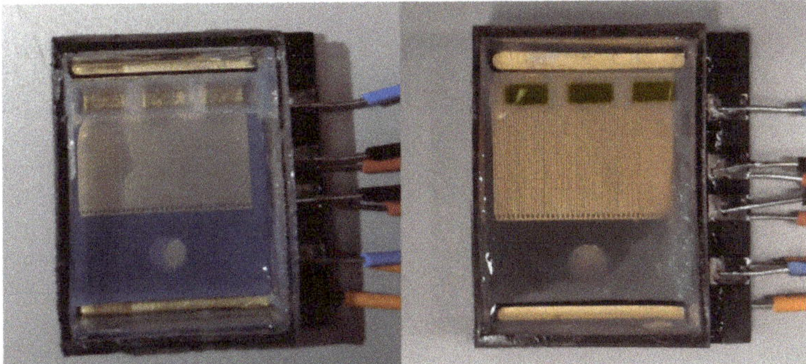

Figure 8. The result of the mixing and curing after removing the structure can be seen. In addition, the wells are filled with liquids.

The device is checked with DNA in order to verify the correct migration of DNA along the agarose gel; Figure 9. This is important to analyse the homogeneity of the agarose gel. In order to do so, the electrophoresis is performed at 40 V with a required current of 20 mA. The negative and positive electrodes are shown in Figure 1 with black and red arrows, respectively.

Figure 9. Bands obtained after migrating the DNA by electrophoresis.

As can be seen, three parallel bands are obtained after the DNA migration by electrophoresis. These bands are not appreciably deformed in the agarose gel. Therefore, the homogeneity is acceptable.

The experimental results show good behaviour of the prototype under the chosen parameters. A balance between bubble generation and rapid migration have to be taken into account. In this respect, the generation of bubbles during the electrophoresis for voltages higher than 50 V is very quick. In this case, the diameter of many generated bubbles is very high, about 2 mm. This is due to the fact that the rapid bubble generation in a small cavity implies that the small bubbles coalesce to form bigger ones. On the other hand, the migration at 40 V is rapid enough—in this case 7 min—so that the gold electrodes support the operation. This migration rate can be improved if the concentration of agarose in the fabricated gel decreases, for example 1%. Regarding the location of the transparent film, it must be placed on the top surface of the PCB. Otherwise, it does not support the temperature of the microheater.

It is worthy to highlight that the central band of the Figure 9 is not deformed after the electrophoresis. However, the two other bands are slightly deformed close to the lateral walls. This effect is due to the fact that the final agarose gel is slightly thicker in that area because the surface tension of the liquid is not negligible. Therefore, the resistance to the migration is higher. We are planning to reduce the dimensions of the wells in future developments of the prototype, in order to increase the distance between them and the walls, and to minimise the deformations.

6. Conclusions

A semi-automatic prototype of a single-use device for agarose gel preparation and electrophoresis is described. It includes a conductivity sensor for detecting the agarose-TAE solution in order to start the automatic process. In addition, the device has an integrated microheater and a NTC thermistor for controlling the mixing temperature. Moreover, an optical sensor is used for measuring the degree of transparency. The device is lab-on-PCB fabricated, using commercial PCB substrates and thermoplastics. This fact implies low cost, fast analysis, low sample consumption, integration capability and mass production.

It is intended as an integrable functional module of a more complex system for DNA amplification and detection, using qualitative PCR, or even a device itself, to perform more controllable electrophoresis in conventional laboratories, minimizing the human factor.

The proposed lab-on-PCB is intended to be integrated with lab on chip thermocyclers and fluorescence detection systems for developing automatic PCR devices. The whole automatic system will include the DNA amplification, agarose gel fabrication, electrophoresis and detection. The application of this device is point-of-care diagnosis based on qualitative PCR. Finally, the proposed system can be used for conventional PCR procedures as an independent lab-on-PCB device. In addition, the device can be fabricated with smaller dimensions for a higher integration on lab on chip, or even larger ones for conventional multiwells agarose gels.

Author Contributions: Conceptualisation, F.P.; methodology, F.P. and J.D.U.-G.; software, F.P.; validation, F.P. and J.D.U.-G.; formal analysis, J.D.U.-G.; investigation, F.P. and J.D.U.-G. ; resources, J.M.Q. and F.P.; data curation, F.P. writing—original draft preparation, F.P. and J.D.U.-G.; writing—review and editing, J.D.U.-G.; visualisation, F.P.; supervision, J.M.Q. and F.P.; project administration, F.P.; funding acquisition, J.M.Q. and F.P. All authors have read and agreed to the published version of the manuscript.

Funding: This work has been funded by regional government Junta de Andalucía (Consejería de Economía y Conocimiento), Plan Andaluz de Investigación, Desarrollo e Innovación (PAIDI 2020) with the project "Sistema para la amplificación y detección de fragmentos de ADN empleando PCR en Lab-on-chip (PCR-on-a-Chip)", reference project P18-RT-1745. The authors also thank "Proyectos I + D + i FEDER Andalucía 2014–2020", project "Lab-on-chip de electro-estimulación, para el estudio In-vitro de Cultivos de retina de Larga duración: Retina-on-a-chip" reference project US-1265983.

Institutional Review Board Statement: Not applicable.

Informed Consent Statement: Not applicable.

Data Availability Statement: Data is contained within the article.

Conflicts of Interest: The authors declare no conflict of interest.

Abbreviations

The following abbreviations are used in this manuscript:

MDPI	Multidisciplinary Digital Publishing Institute
NTC	Negative Temperature Coefficient
LDR	Light Dependent Resistor
PCR	Polymerase Chain Reaction
PCB	Printed Circuit Board

References

1. Perdigones, F. Lab-on-PCB and Flow Driving: A Critical Review. *Micromachines* **2021**, *12*, 175. [CrossRef]
2. Moschou, D.; Tserepi, A. The lab-on-PCB approach: Tackling the μTAS commercial upscaling bottleneck. *Lab Chip* **2017**, *17*, 1388–1405. [CrossRef] [PubMed]
3. Zhao, W.; Tian, S.; Huang, L.; Liu, K.; Dong, L. The review of Lab-on-PCB for biomedical application. *Electrophoresis* **2020**, *41*, 1433–1445. [CrossRef] [PubMed]
4. Flores, G.; Aracil, C.; Perdigones, F.; Quero, J.M. Lab-protocol-on-PCB: Prototype of a laboratory protocol on printed circuit board using MEMS technologies. *Microelectron. Eng.* **2018**, *200*, 26–31. [CrossRef]
5. Chang, Y.; You, H. Efficient Bond of PDMS and Printed Circuit Board with An Application on Continuous-flow Polymerase Chain Reaction. *BioChip J.* **2020**, *14*, 349–357. [CrossRef]
6. Burdallo, I.; Jimenez-Jorquera, C.; Fernández-Sánchez, C.; Baldi, A. Integration of microelectronic chips in microfluidic systems on printed circuit board. *J. Micromech. Microeng.* **2012**, *22*, 105022. [CrossRef]
7. Nikshoar, M.S.; Khosravi, S.; Jahangiri, M.; Zandi, A.; Miripour, Z.S.; Bonakdar, S.; Abdolahad, M. Distinguishment of populated metastatic cancer cells from primary ones based on their invasion to endothelial barrier by biosensor arrays fabricated on nanoroughened poly (methyl methacrylate). *Biosens. Bioelectron.* **2018**, *118*, 51–57. [CrossRef]
8. Jolly, P.; Rainbow, J.; Regoutz, A.; Estrela, P.; Moschou, D. A PNA-based Lab-on-PCB diagnostic platform for rapid and high sensitivity DNA quantification. *Biosens. Bioelectron.* **2019**, *123*, 244–250. [CrossRef]
9. Cabello, M.; Mozo, M.; De la Cerda, B.; Aracil, C.; Diaz-Corrales, F.J.; Perdigones, F.; Valdes-Sanchez, L.; Relimpio, I.; Bhattacharya, S.S.; Quero, J.M. Electrostimulation in an autonomous culture lab-on-chip provides neuroprotection of a retinal explant from a retinitis pigmentosa mouse-model. *Sens. Actuators B Chem.* **2019**, *288*, 337–346. [CrossRef]
10. Anastasova, S.; Kassanos, P.; Yang, G.Z. Multi-parametric rigid and flexible, low-cost, disposable sensing platforms for biomedical applications. *Biosens. Bioelectron.* **2018**, *102*, 668–675. [CrossRef]
11. Voeten, R.L.; Ventouri, I.K.; Haselberg, R.; Somsen, G.W. Capillary electrophoresis: Trends and recent advances. *Anal. Chem.* **2018**, *90*, 1464–1481. [CrossRef]
12. Mantovani, V.; Galeotti, F.; Maccari, F.; Volpi, N. Recent advances in capillary electrophoresis separation of monosaccharides, oligosaccharides, and polysaccharides. *Electrophoresis* **2018**, *39*, 179–189. [CrossRef]
13. Hajba, L.; Guttman, A. Recent advances in column coatings for capillary electrophoresis of proteins. *TrAC Trends Anal. Chem.* **2017**, *90*, 38–44. [CrossRef]
14. Toraño, J.S.; Ramautar, R.; de Jong, G. Advances in capillary electrophoresis for the life sciences. *J. Chromatogr. B* **2019**, *1118*, 116–136. [CrossRef]
15. De Carvalho, J.A.; Pinheiro, B.G.; Hagen, F.; Gonçalves, S.S.; Negroni, R.; Kano, R.; Bonifaz, A.; de Camargo, Z.P.; Rodrigues, A.M. A new duplex PCR assay for the rapid screening of mating-type idiomorphs of pathogenic Sporothrix species. *Fungal Biol.* **2021**. [CrossRef]
16. Kaprou, G.D.; Papadopoulos, V.; Loukas, C.M.; Kokkoris, G.; Tserepi, A. Towards PCB-based miniaturized thermocyclers for DNA amplification. *Micromachines* **2020**, *11*, 258. [CrossRef] [PubMed]
17. Kaprou, G.D.; Papadopoulos, V.; Papageorgiou, D.P.; Kefala, I.; Papadakis, G.; Gizeli, E.; Chatzandroulis, S.; Kokkoris, G.; Tserepi, A. Ultrafast, low-power, PCB manufacturable, continuous-flow microdevice for DNA amplification. *Anal. Bioanal. Chem.* **2019**, *411*, 5297–5307. [CrossRef] [PubMed]
18. Lee, P.Y.; Saraygord-Afshari, N.; Low, T.Y. The evolution of two-dimensional gel electrophoresis-from proteomics to emerging alternative applications. *J. Chromatogr. A* **2020**, *1615*, 460763. [CrossRef] [PubMed]
19. Minipcr Bio^{TM}. Available online: https://www.minipcr.com/wp-content/uploads/Three-ways-to-pour-agarose-gels_final.pdf (accessed on 18 August 2016).

20. Merck KGaA. Available online: https://www.sigmaaldrich.com/deepweb/assets/sigmaaldrich/product/documents/181/864/a9539pis.pdf (accessed on 18 August 2016).
21. Invitrogen^{TM}. Available online: https://www.thermofisher.com/order/catalog/product/16500500#/16500500 (accessed on 18 August 2016).
22. Strobel, H.A.; Calamari, E.L.; Alphonse, B.; Hookway, T.A.; Rolle, M.W. Fabrication of custom agarose wells for cell seeding and tissue ring self-assembly using 3D-printed molds. *JoVE (J. Vis. Exp.)* **2018**, *134*, e56618. [CrossRef]
23. Zhang, C.; Xu, J.; Ma, W.; Zheng, W. PCR microfluidic devices for DNA amplification. *Biotechnol. Adv.* **2006**, *24*, 243–284. [CrossRef]
24. Zhang, Y.; Ozdemir, P. Microfluidic DNA amplification—A review. *Anal. Chim. Acta* **2009**, *638*, 115–125. [CrossRef]
25. Kulkarni, M.B.; Goel, S. Advances in continuous-flow based microfluidic PCR devices—A review. *Eng. Res. Express* **2020**, *2*, 042001.
26. Zhang, C.X.; Manz, A. High-speed free-flow electrophoresis on chip. *Anal. Chem.* **2003**, *75*, 5759–5766. [CrossRef]
27. Ou, X.; Chen, P.; Huang, X.; Li, S.; Liu, B.F. Microfluidic chip electrophoresis for biochemical analysis. *J. Sep. Sci.* **2020**, *43*, 258–270. [CrossRef] [PubMed]
28. Jung, Y.K.; Kim, J.; Mathies, R.A. Microfluidic linear hydrogel array for multiplexed single nucleotide polymorphism (SNP) detection. *Anal. Chem.* **2015**, *87*, 3165–3170. [CrossRef]
29. He, M.; Herr, A.E. Automated microfluidic protein immunoblotting. *Nat. Protoc.* **2010**, *5*, 1844–1856. [CrossRef] [PubMed]
30. Vlassakis, J.; Herr, A.E. Joule heating-induced dispersion in open microfluidic electrophoretic cytometry. *Anal. Chem.* **2017**, *89*, 12787–12796. [CrossRef] [PubMed]
31. Pan, Q.; Yamauchi, K.A.; Herr, A.E. Controlling dispersion during single-cell polyacrylamide-gel electrophoresis in open microfluidic devices. *Anal. Chem.* **2018**, *90*, 13419–13426. [CrossRef] [PubMed]
32. Duncombe, T.A.; Herr, A.E. Photopatterned free-standing polyacrylamide gels for microfluidic protein electrophoresis. *Lab Chip* **2013**, *13*, 2115–2123. [CrossRef]
33. Tentori, A.M.; Hughes, A.J.; Herr, A.E. Microchamber integration unifies distinct separation modes for two-dimensional electrophoresis. *Anal. Chem.* **2013**, *85*, 4538–4545. [CrossRef]
34. Li, Y.; Feng, X.; Du, W.; Li, Y.; Liu, B.F. Ultrahigh-throughput approach for analyzing single-cell genomic damage with an agarose-based microfluidic comet array. *Anal. Chem.* **2013**, *85*, 4066–4073. [CrossRef] [PubMed]

MDPI
St. Alban-Anlage 66
4052 Basel
Switzerland
Tel. +41 61 683 77 34
Fax +41 61 302 89 18
www.mdpi.com

Micromachines Editorial Office
E-mail: micromachines@mdpi.com
www.mdpi.com/journal/micromachines

www.ingramcontent.com/pod-product-compliance
Lightning Source LLC
LaVergne TN
LVHW070849170726
843515LV00005B/607
9783036547022